PROBLEMS FOR ENGINEERING GRAPHICS COMMUNICATION AND TECHNICAL GRAPHICS COMMUNICATION

Revised Edition

Prepared by

The Faculty at Purdue University
Technical Graphics Department

Edited by
Gary R. Bertoline

IRWIN

Chicago • Bogotá • Boston • Buenos Aires • Caracas
London • Madrid • Mexico City • Sydney • Toronto

PROBLEMS FOR ENGINEERING GRAPHICS COMMUNICATION AND TECHNICAL GRAPHICS COMMUNICATION Revised Edition

1 2 3 4 5 6 7 8 9 0 QD QD 9 0 9 8 7
ISBN 0–256–26780–4

http://www.mhcollege.com

THE IRWIN GRAPHICS SERIES

Titles in the Irwin Graphics Series include:

Technical Graphics Communication, 2nd Edition
Bertoline, Wiebe, Miller, Mohler, 1996

Technical Graphic Communication, 1st Edition
Bertoline, Wiebe, Miller, Nasman, 1995

Fundamentals of Graphics Communication
Bertoline, Wiebe, Miller, Nasman, 1996

Engineering Graphics Communication
Bertoline, Wiebe, Miller, Nasman, 1995

Graphics Interactive CD-Rom
Dennis Lieu, 1997

Problems for Engineering Graphics Communication and Technical Graphics Communication, Revised 1st Edition, 1997

AutoCAD Instructor 13
James A. Leach, 1996

AutoCAD Companion 13
James A. Leach, 1996

AutoCAD Instructor 12
James A. Leach, 1995

AutoCAD Companion 12
James A. Leach, 1995

The CADKEY Companion
John Cherng, 1995

Hands-On CADKEY
Timothy Sexton, 1995

Engineering Design Visualization Workbook
Dennis Stevenson, 1997

Modeling with AutoCAD Designer
Sandra Dobek and Ryan Ranschaert, 1996

Providing you with the highest quality textbooks that meet your changing needs requires feedback, improvement, and revision. The team of authors and Richard D. Irwin, Inc., are committed to this effort. We invite you to become part of our team by offering your wishes, suggestions, and comments for future editions and new products and texts.

Please mail or fax your comments to:

Gary Bertoline
c/o Richard D. Irwin, Inc.
1333 Burr Ridge Parkway
Burr Ridge, IL 60521
fax: 708–789–6946

Preface

Problems for Engineering Graphics Communication and Technical Graphics Communication, have been designed as a practical supplement for *Engineering Graphics Communication* and *Technical Graphics Communication.* But these problem books have been designed to be used with any engineering or technical graphics textbook.

Instructions are given with each page, but further instructions may need to be given by the instructor for individual problems. Many of the problems can be solved by sketching or through of computer-aided-design (CAD) or they can be solved by traditional methods.

The problems have been designed so that they make the students reference the textbook and the problems force the students to use problem solving techniques to determine the solution. The problems are designed to cover the basic principles of engineering and technical graphics but the workbook is not designed as a stand alone book. The workbooks include problems for sectioning, dimensioning, threaded fasteners, multiviews, sketching, isometrics, geometric construction, auxiliary views, and descriptive geometry.

These problems have been developed and used in the Department of Technical Graphics by graphics professionals over the last fifty years. Thousands of students all over the state of Indiana have used the problems over the years. Many individuals have made contributions to this workbook. This workbook is dedicated to these individuals and any royalties earned will be used by The Department of Technical Graphics to further develop and teach all young people the importance of technical graphics.

CONTENTS

LETTERING

CAPITAL LETTERS

INSTRUCTIONS: The grids below are identified for each letter and numeral. The height of each character equals the height of its grid and, except in the case of the two letters F and J, the width of each character is two units less than the width of its grid. The position of each curved outline is indicated by a series of small dots.

On the grids, using a pencil having a lead properly shaped and pointed, make each symbol on the appropriate grid. Follow the order of stroke and direction of stroke as given in the text reading references and related illustrations. In making each character look for possible related similarity to other characters as well as for the differences and pecularities.

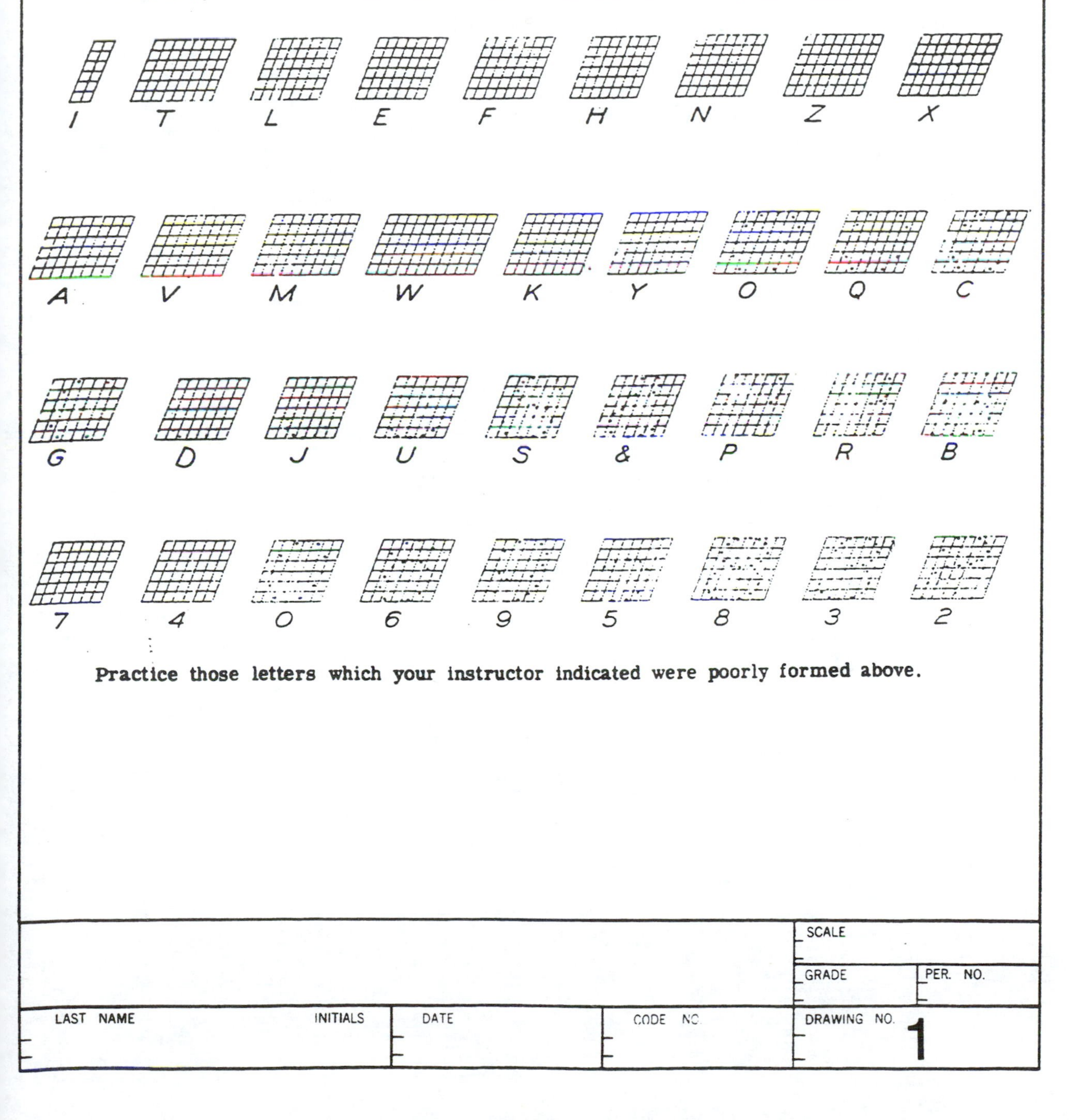

Practice those letters which your instructor indicated were poorly formed above.

GEOMETRY REVIEW

INSTRUCTIONS: From this sheet the student is expected to determine how well he can recall identifying terms that are associated with geometric shapes. These are the terms with which one should have become familiar while in high school.

Using the guide lines provided, letter the single word that provides the best identification for what has been shown. For example, the one word providing a limited description for the angle shown in (1) is the word RIGHT. Good lettering is expected.

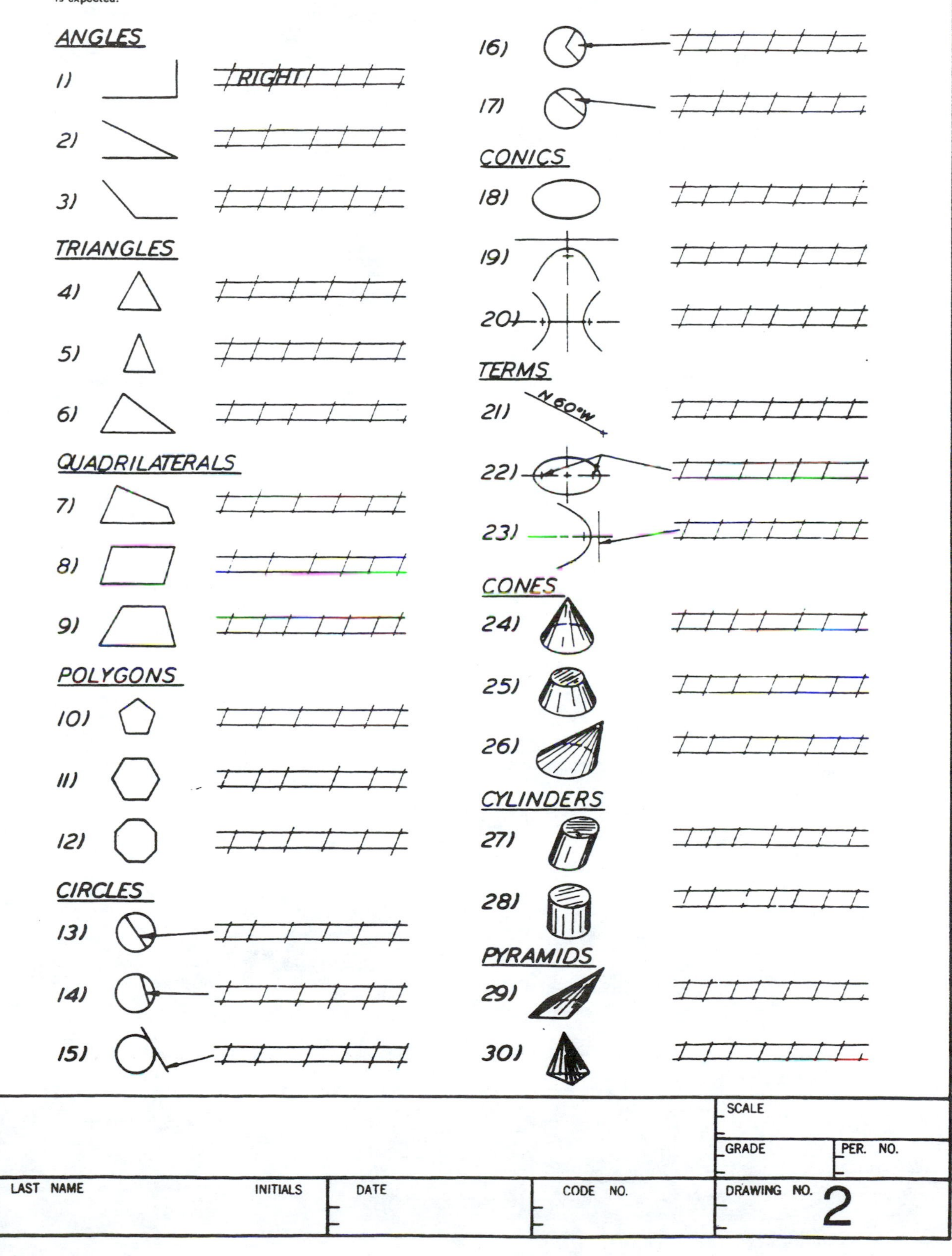

SCALES

USE OF THE ENGINEERS' (DECIMAL) SCALE AND INSTRUMENTS

INTRODUCTIONS: All engineers must be proficient in the use of scales. The scale reading problems on this sheet are intended to familiarize the student with the commonly used decimal scales Since all of the decimal scales are used similarly, the only difference being the number of divisions per inch (10, 20, 30, 40, 50) an understanding of the use of the scale marked FULL SIZE should make it possible for one to read and use all of the other decimal scales. The division on the various scales may represent different units of measurement. In using the scale marked FULL SIZE where each inch is divided into fifty parts, the one-tenth inch divisions are distinguishable and represent one-tenth of an inch, or, when representing very large objects or long ground distances on a drawing, each inch could represent 100 feet and the tenth-inch divisions ten feet.

Examples:

USING SCALE MARKED FULL SIZE	USING SCALE MARKED ¼ SIZE
If 1 in = 1 ft, then the smallest unit equals .02 ft.	If 1 in = 4 ft, then the smallest unit equals .1 ft
If 1 in = 10 ft, then the smallest unit equals .2 ft.	If 1 in = 40 ft, then the smallest unit equals 1 ft
If 1 in = 100 ft, then the smallest unit equals 2 ft.	If 1 in = 400 ft, then the smallest unit equals 10 ft

Prob. 1. Using the scale indicated, scale and record the length represented by each of the given lines to two decimal places wherever appropriate. Show "(inch) and '(foot) marks.

USE SCALE MARKED FULL SIZE

Full Size (1 in = 1 in)

1 in = 10 ft

1 in = 100 ft

USE 40 SCALE

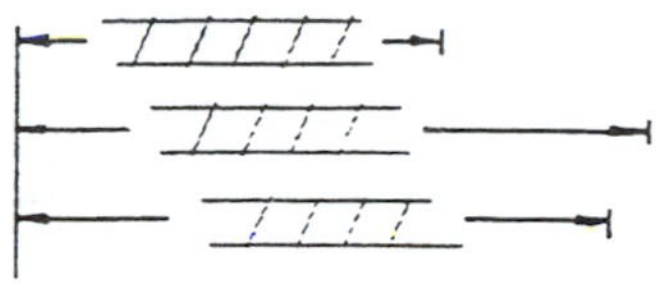

Quarter Size 1 in. = 4.00 in.

1 in = 40 ft

1 in = 4000 ft

USE 50 SCALE

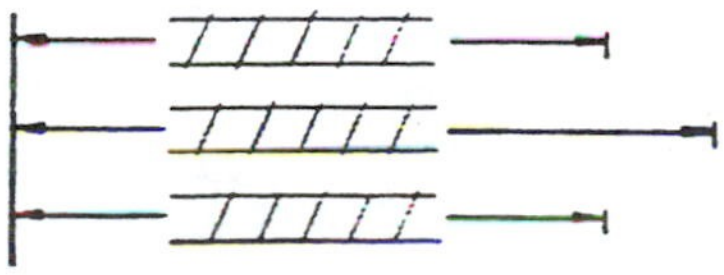

1 in = 5 in

1 in = 500 ft

1 in = 50 ft

USE 20 SCALE

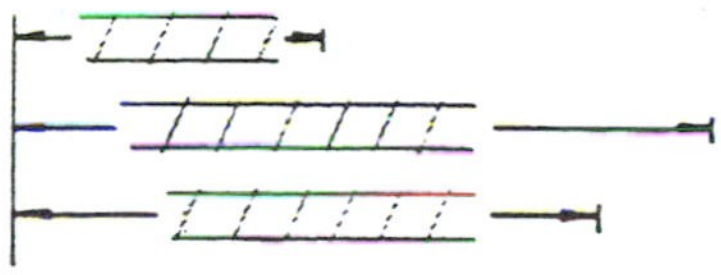

Half Size 1 in. = 2.00 in.

1 in = 2000 ft.

1 in = 20 ft

Prob. 2. How would the scale be specified on a drawing for each of the given lines. Record these scale statements using the guide lines that have been provided.

SCALE

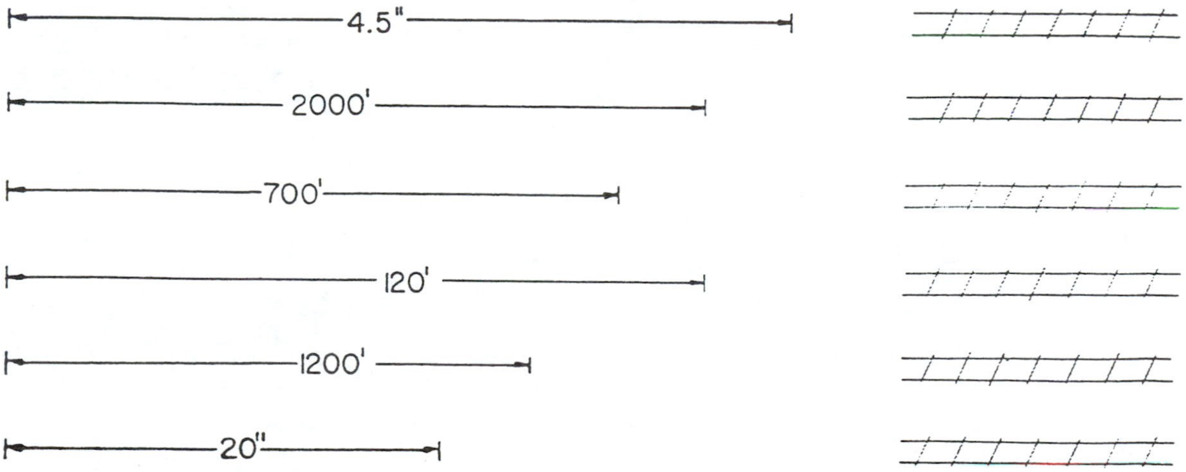

4.5"

2000'

700'

120'

1200'

20"

	SCALE	
	GRADE	PER. NO.
LAST NAME / INITIALS / DATE / CODE NO.	DRAWING NO.	3

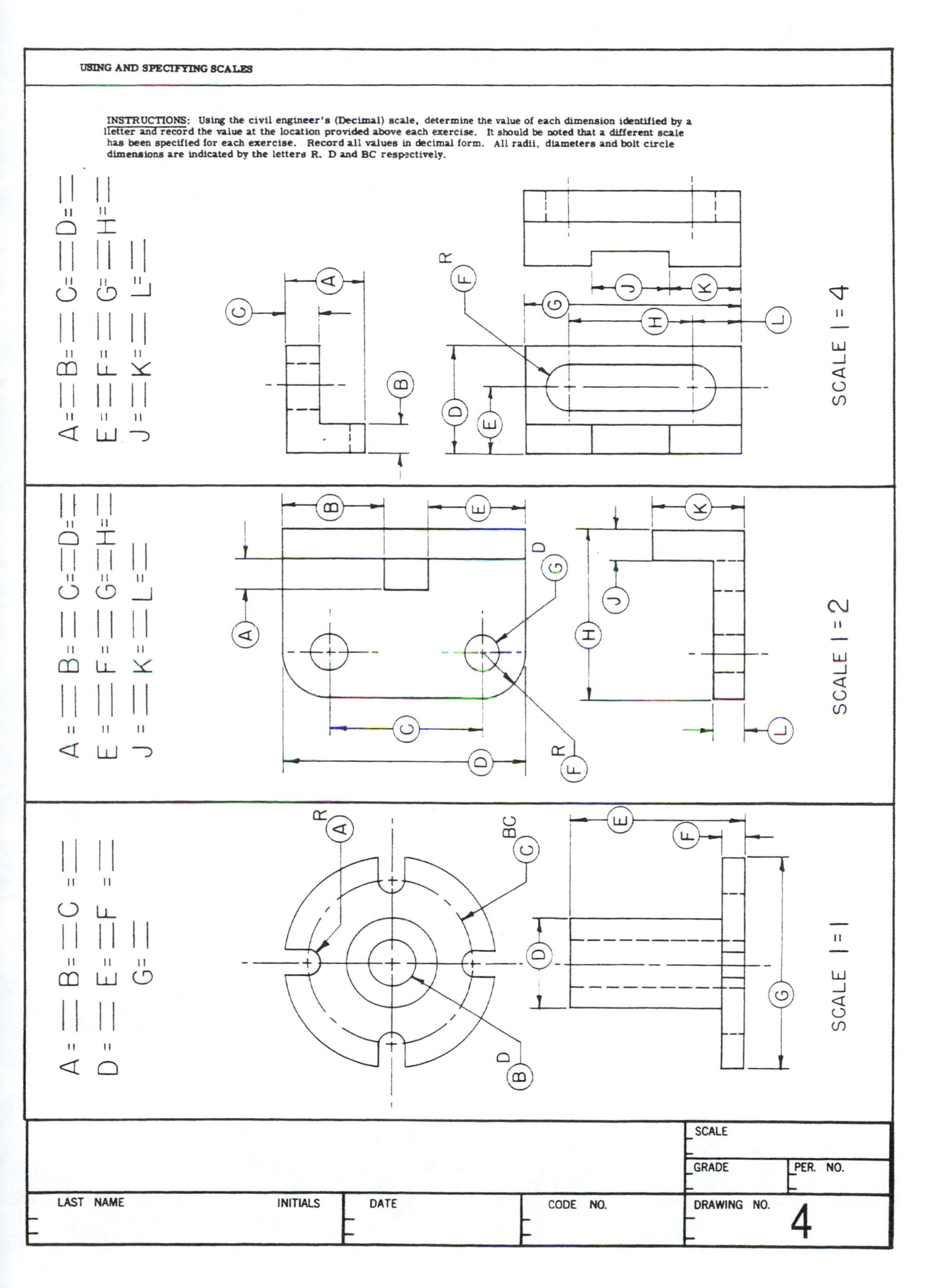

USING AND SPECIFYING SCALES
INSTRUCTIONS: Using the civil engineer's (Decimal) scale, determine the value of each dimension identified by a letter and record the value at the location provided above each exercise. It should be noted that a different scale has been specified for each exercise. Record all values in decimal form. All radii, diameters and bolt circle dimensions are indicated by the letters R, D and BC respectively.
A= B= C= D=
E= F= G= H=
J= K= L=
SCALE 1 = 4
A= B= C= D=
E= F= G= H=
J= K= L=
SCALE 1 = 2
A= B= C=
D= E= F=
G=
SCALE 1 = 1
SCALE
GRADE
PER. NO.
LAST NAME
INITIALS
DATE
CODE NO.
DRAWING NO.
4

These problems are to be solved using the T-square, triangles, scale and dividers as appropriate.

Prob. 3. A control panel is to have 8 toggle switches placed in a horizontal row. For appearance they are to be equally spaced, with the end switches located as shown. Locate the remaining 6 switches.

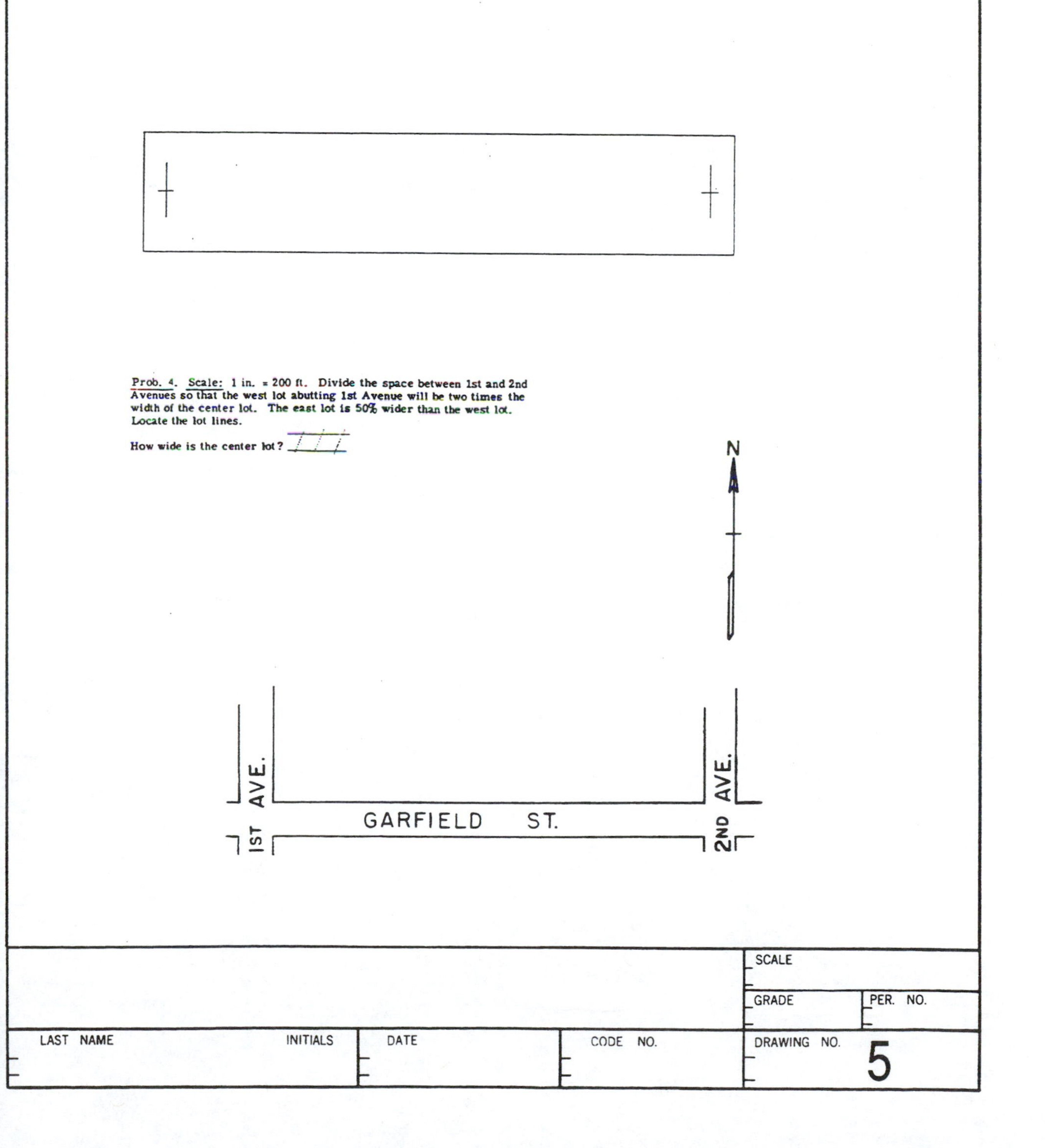

Prob. 4. Scale: 1 in. = 200 ft. Divide the space between 1st and 2nd Avenues so that the west lot abutting 1st Avenue will be two times the width of the center lot. The east lot is 50% wider than the west lot. Locate the lot lines.

How wide is the center lot? ________

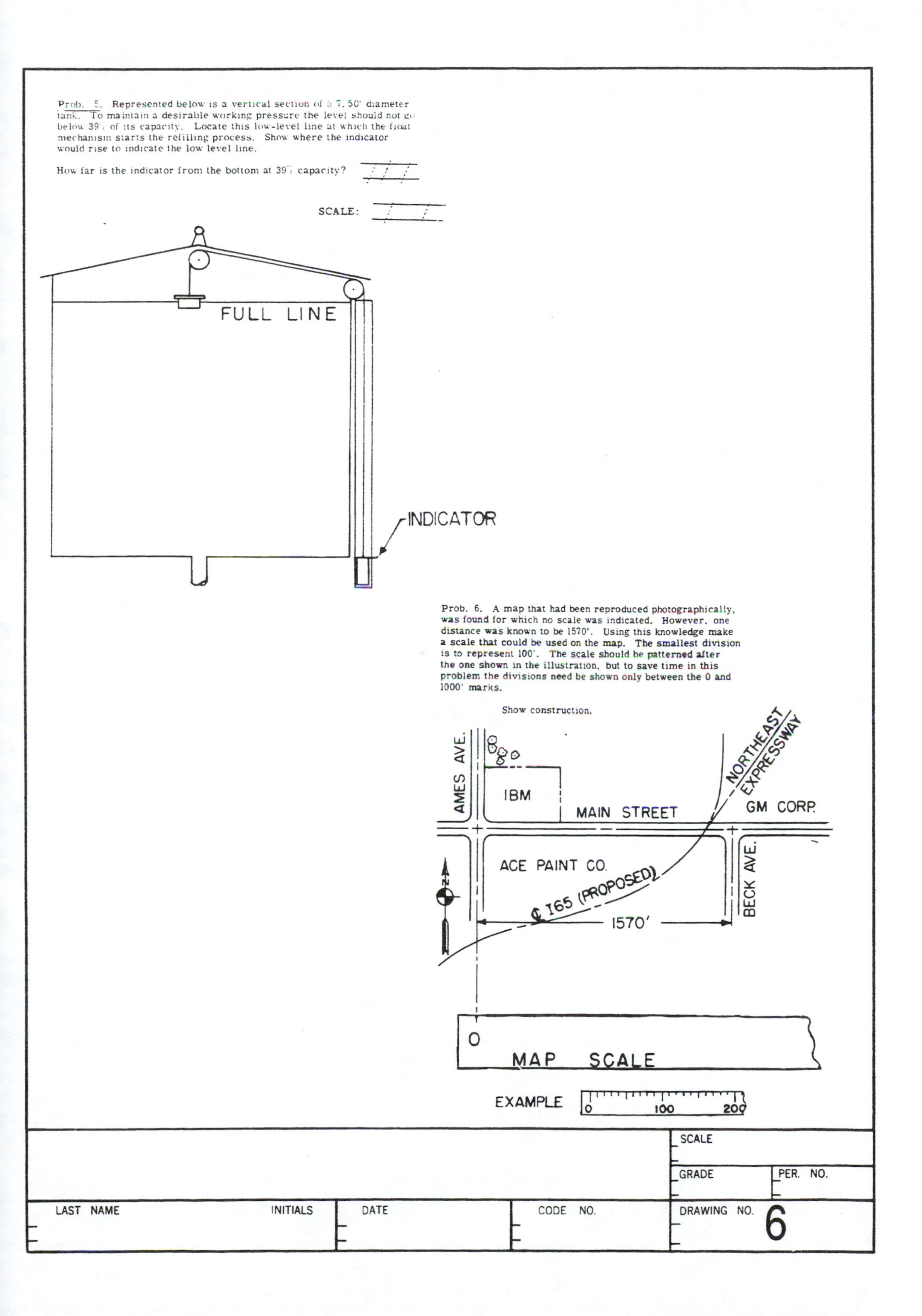
Prob. 5. Represented below is a vertical section of a 7.50' diameter tank. To maintain a desirable working pressure the level should not go below 39% of its capacity. Locate this low-level line at which the float mechanism starts the refilling process. Show where the indicator would rise to indicate the low level line.
How far is the indicator from the bottom at 39% capacity?
SCALE:
FULL LINE
INDICATOR
Prob. 6. A map that had been reproduced photographically, was found for which no scale was indicated. However, one distance was known to be 1570'. Using this knowledge make a scale that could be used on the map. The smallest division is to represent 100'. The scale should be patterned after the one shown in the illustration, but to save time in this problem the divisions need be shown only between the 0 and 1000' marks.
Show construction.
AMES AVE.
IBM
MAIN STREET
NORTHEAST EXPRESSWAY
GM CORP.
ACE PAINT CO.
℄ I65 (PROPOSED)
BECK AVE.
1570'
N
0
MAP SCALE
EXAMPLE
0
100
200
SCALE
GRADE
PER. NO.
LAST NAME
INITIALS
DATE
CODE NO.
DRAWING NO.
6

GEOMETRIC CONSTRUCTIONS

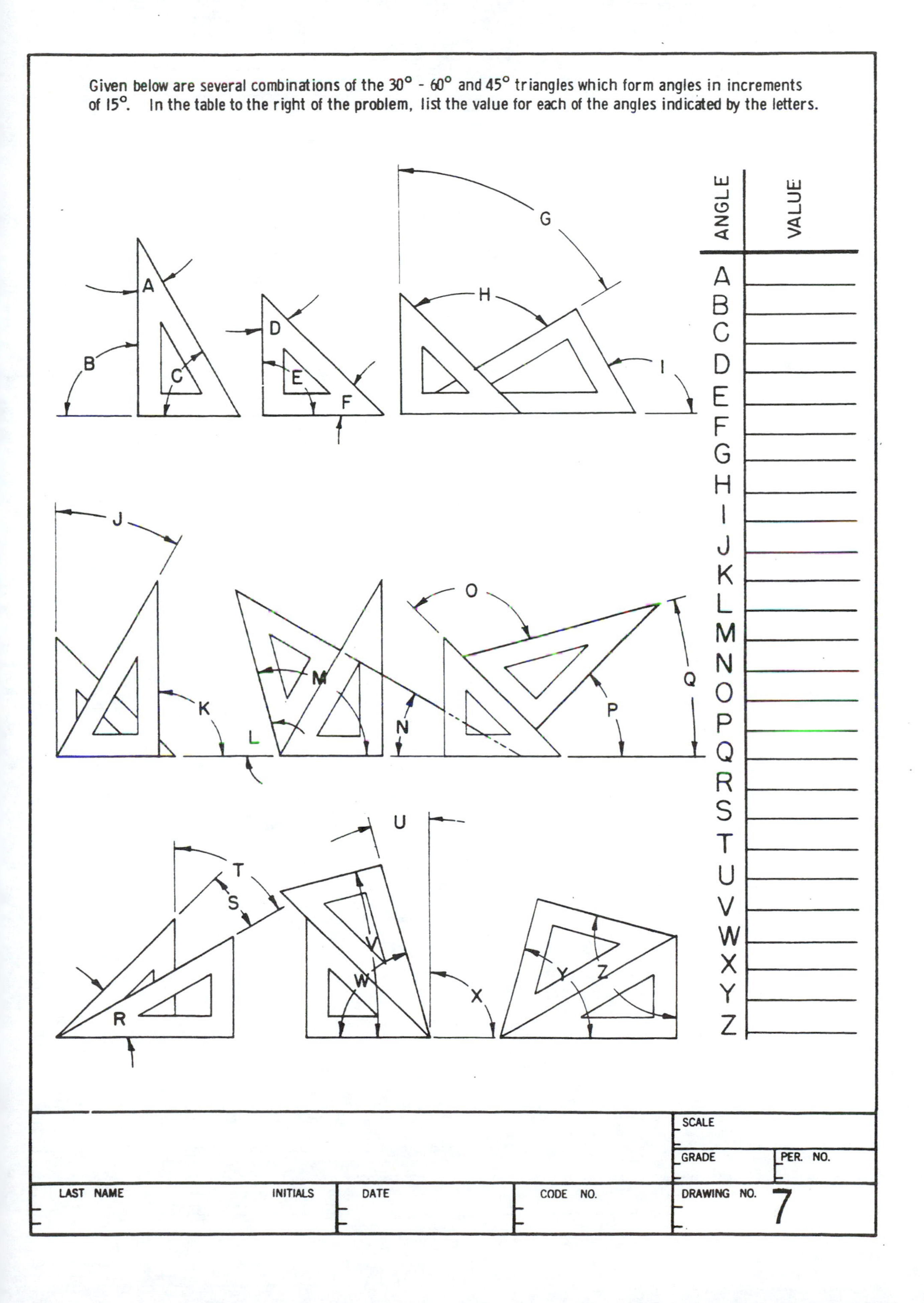

Given below are several combinations of the 30° - 60° and 45° triangles which form angles in increments of 15°. In the table to the right of the problem, list the value for each of the angles indicated by the letters.

ANGLE	VALUE
A	
B	
C	
D	
E	
F	
G	
H	
I	
J	
K	
L	
M	
N	
O	
P	
Q	
R	
S	
T	
U	
V	
W	
X	
Y	
Z	

			SCALE	
			GRADE	PER. NO.
LAST NAME INITIALS	DATE	CODE NO.	DRAWING NO. 7	

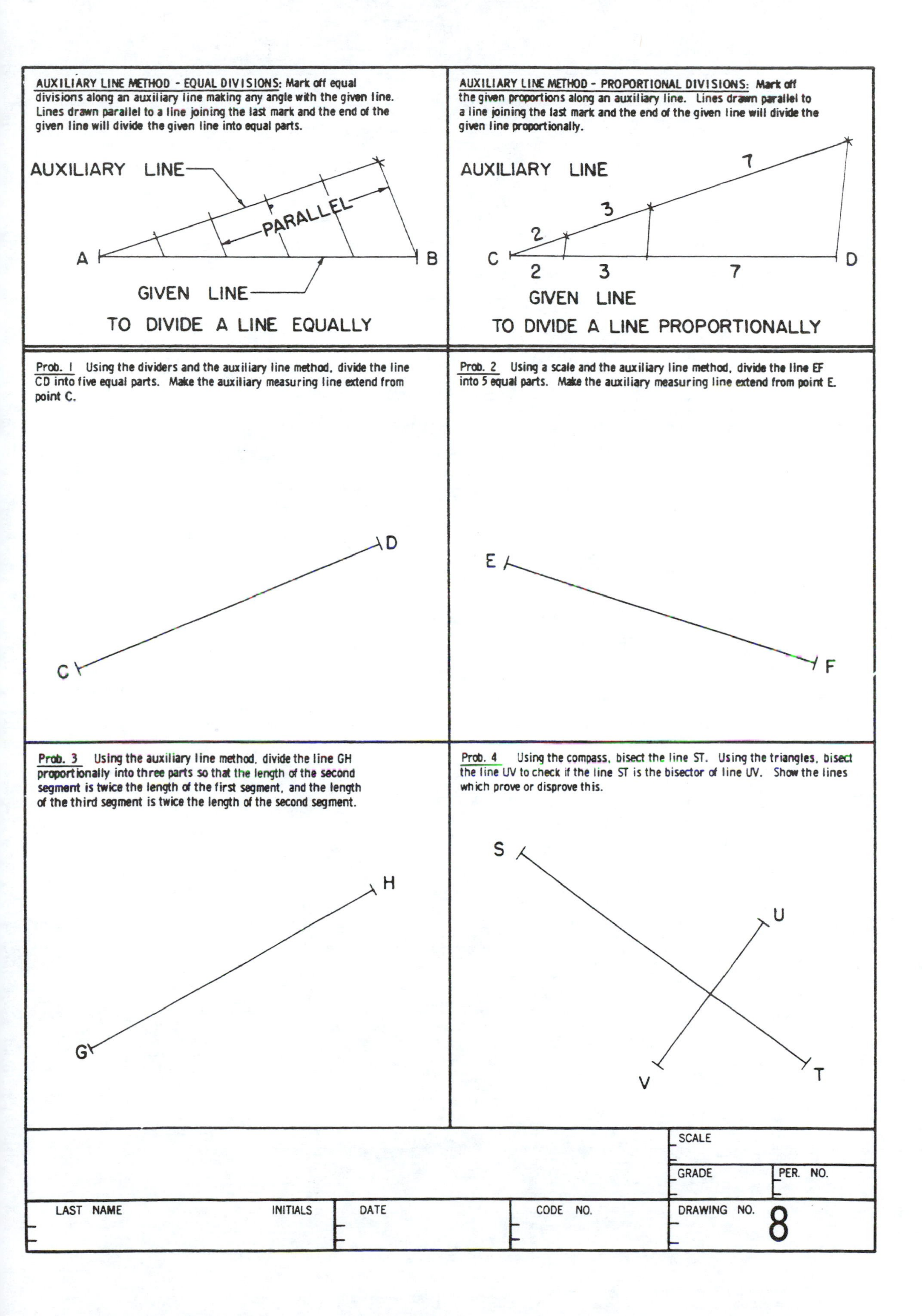
AUXILIARY LINE METHOD - EQUAL DIVISIONS: Mark off equal divisions along an auxiliary line making any angle with the given line. Lines drawn parallel to a line joining the last mark and the end of the given line will divide the given line into equal parts.
AUXILIARY LINE
PARALLEL
A
B
GIVEN LINE
TO DIVIDE A LINE EQUALLY
AUXILIARY LINE METHOD - PROPORTIONAL DIVISIONS: Mark off the given proportions along an auxiliary line. Lines drawn parallel to a line joining the last mark and the end of the given line will divide the given line proportionally.
AUXILIARY LINE
7
3
2
C
D
2
3
7
GIVEN LINE
TO DIVIDE A LINE PROPORTIONALLY
Prob. 1 Using the dividers and the auxiliary line method, divide the line CD into five equal parts. Make the auxiliary measuring line extend from point C.
D
C
Prob. 2 Using a scale and the auxiliary line method, divide the line EF into 5 equal parts. Make the auxiliary measuring line extend from point E.
E
F
Prob. 3 Using the auxiliary line method, divide the line GH proportionally into three parts so that the length of the second segment is twice the length of the first segment, and the length of the third segment is twice the length of the second segment.
H
G
Prob. 4 Using the compass, bisect the line ST. Using the triangles, bisect the line UV to check if the line ST is the bisector of line UV. Show the lines which prove or disprove this.
S
U
V
T
SCALE
GRADE
PER. NO.
LAST NAME
INITIALS
DATE
CODE NO.
DRAWING NO.
8

INSTRUCTIONS: (1) Draw 4 arcs of 14.30 R tangent to the circle A and to the circle defined by points 1, 2, and 3. The four centers are to lie to the left of the vertical centerline through A. Circle A has a 4.90R. (2) Draw an arc of 10.20 R through the center of the circle that has points 1, 2, and 3, and tangent to the arc that lies farthest to the right. Mark each tangent point with a short mark normal to the line.

SCALE: To be determined from the information given.

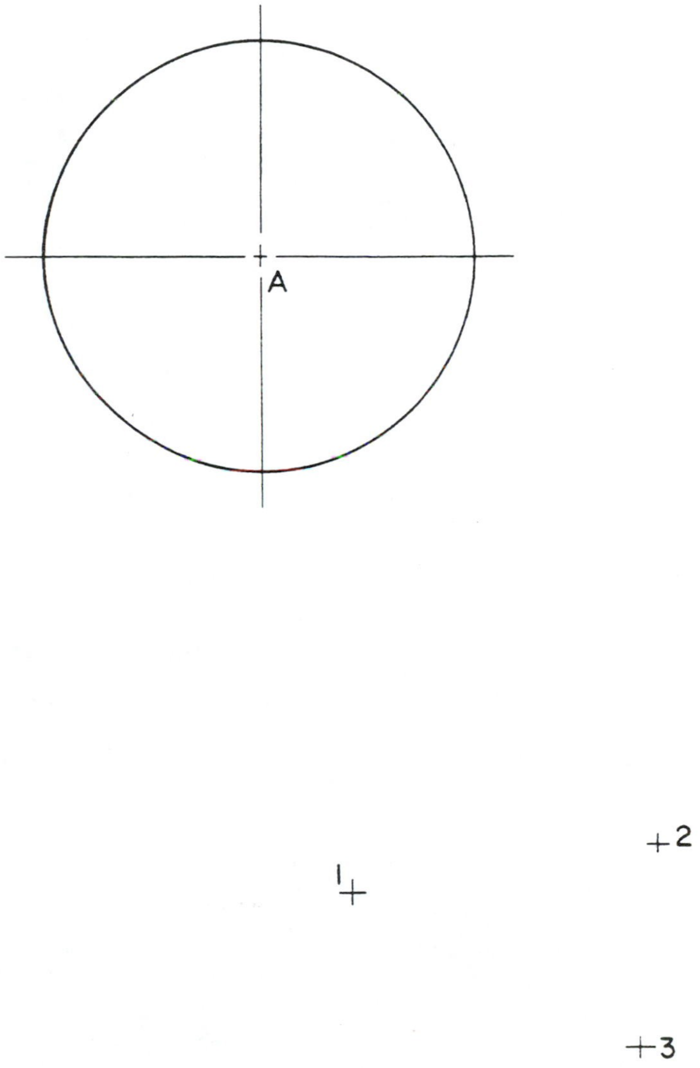

A

2

1

SCALE: ______

3

	SCALE	
	GRADE	PER. NO.

LAST NAME INITIALS	DATE	CODE NO.	DRAWING NO. 9

Mark all tangent points and centers. Determine tangent points before drawing arcs. All arcs are to start and stop at tangent points.

Prob. 1. Complete the front view of a 40 watt electric lamp. The top of the lamp is a portion of a sphere, with a radius of 1.12, tangent to the conical portion shown. Use only an approved geometrical method to find the location of the center of the arc. A trial and error method is not acceptable. Construction lines should be light and thin; finished object lines should be of medium width and very black. Tangent points are to be established before final drawing of the arc and are to be marked.

SCALE: Full Size

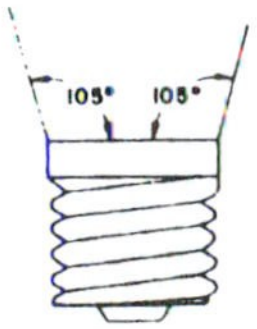

105° 105°

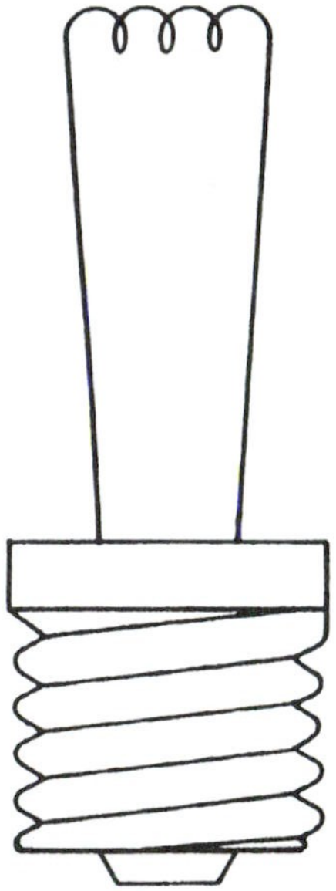

Prob. 2. Complete the front view of the sunlamp. A middle section through the lamp shows the lens portion to be a circular arc of 3.58 radius with its center in the lamp base as shown. The straight sides of the conical reflector are blended into the lens portion with .55 R (radius) arcs. Mark all points of tangency and the centers of the arcs.

SCALE: Full Size

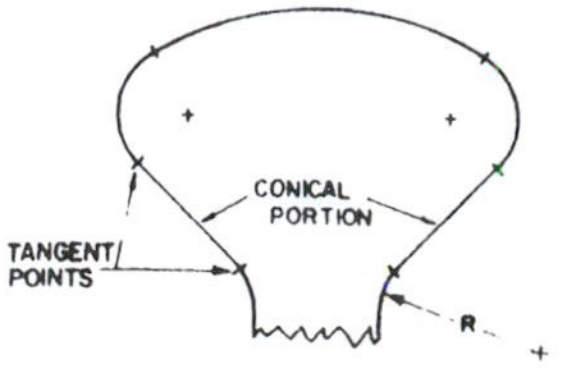

CONICAL PORTION

TANGENT POINTS

R

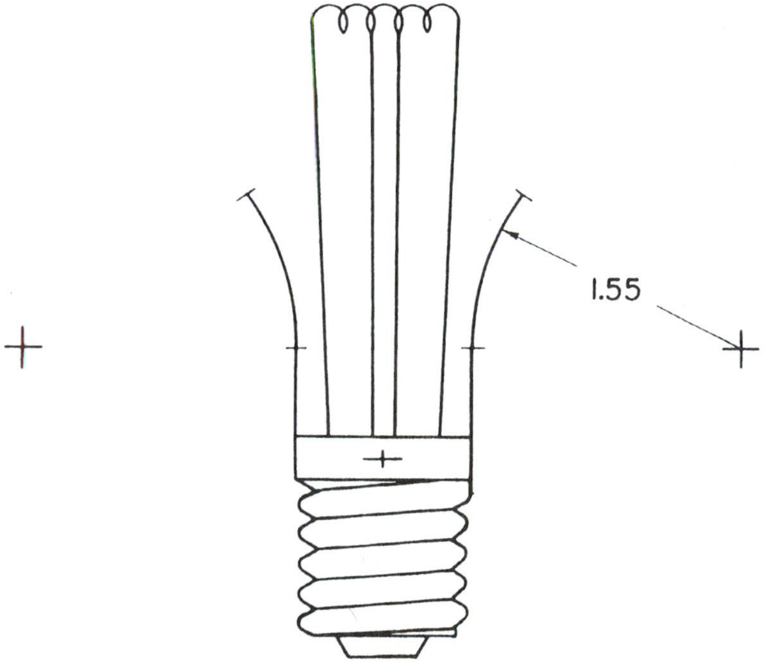

1.55

LAST NAME	INITIALS	DATE	CODE NO.	DRAWING NO. 10

SCALE	
GRADE	PER. NO.

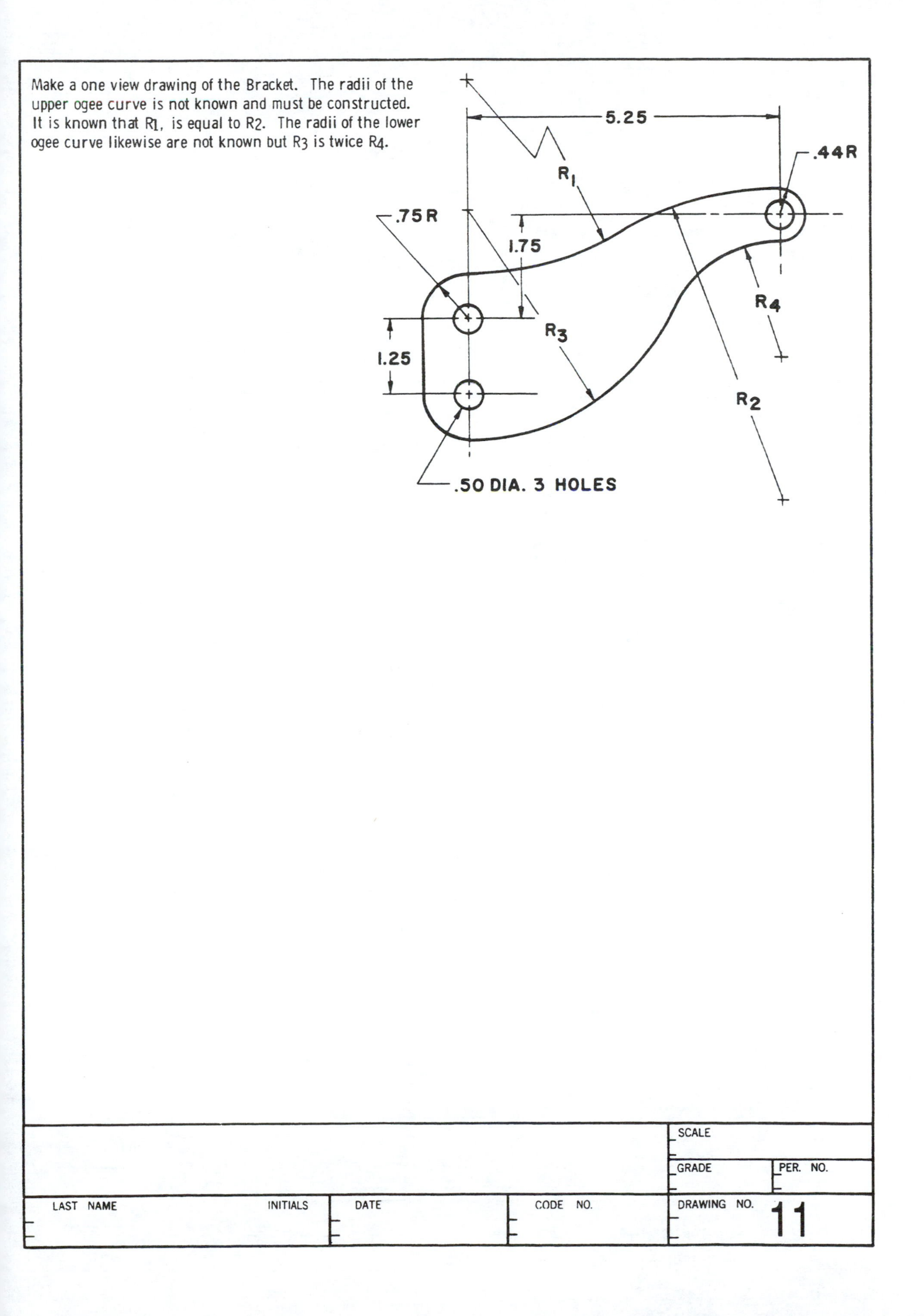
Make a one view drawing of the Bracket. The radii of the upper ogee curve is not known and must be constructed. It is known that R1, is equal to R2. The radii of the lower ogee curve likewise are not known but R3 is twice R4.
5.25
.44R
R1
.75R
1.75
R4
R3
1.25
R2
.50 DIA. 3 HOLES
SCALE
GRADE
PER. NO.
LAST NAME
INITIALS
DATE
CODE NO.
DRAWING NO.
11

Utilizing the information on the accompanying view, make a full size drawing of the object involving the following geometrical principles. Construction of:

a) Arcs tangent to arcs
b) Arcs tangent to lines
c) Angles by the tangent method
d) A triangle given two angles and the included side
e) A square given one side

Use a properly sharpened pencil of medium - hard grade for all construction. Draw finished lines to the widths recommended for each type of line, using a medium grade pencil correctly pointed to give the width needed for the particular type of line being drawn.

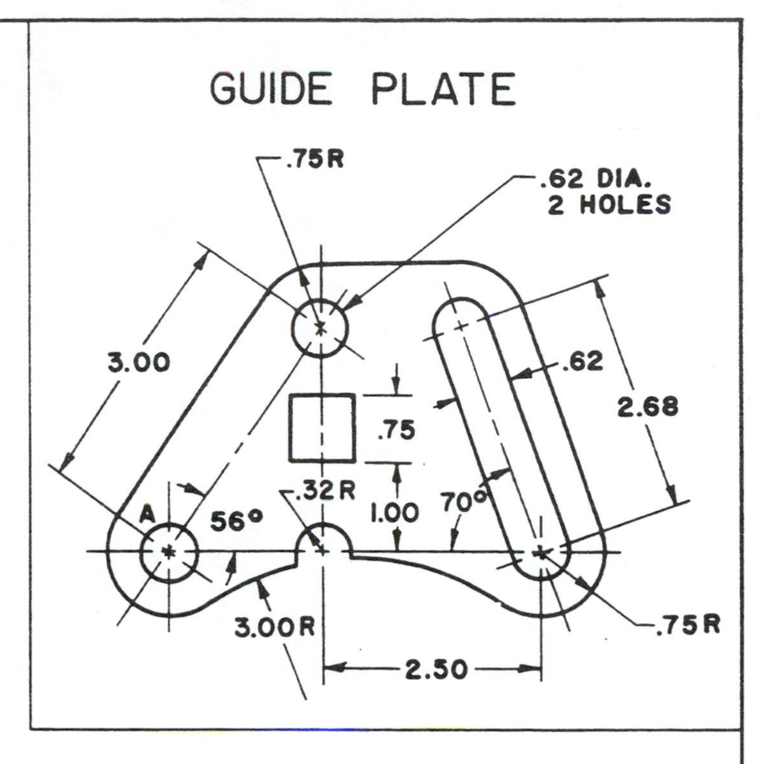

LAST NAME	INITIALS	DATE	CODE NO.	SCALE
				GRADE / PER. NO.
				DRAWING NO. 12

GEOMETRY - ARCS, ANGLES AND POLYGONS

Utilizing the information on the accompanying view, make a full size drawing of the object involving the following geometrical principles. Construction of:

a) Arcs tangent to arcs
b) Arcs tangent to inclined lines
c) Angles by the tangent method
d) A triangle given three sides
e) A triangle given two sides and the included angle
f) A regular hexagon given the length of one side

Use a properly sharpened pencil of medium-hard grade for all construction. Draw finished lines to the widths recommended for each type of line, using a medium grade pencil correctly pointed to give the width needed for the particular type of line being drawn. Omit phantom and dimension lines on final drawing.

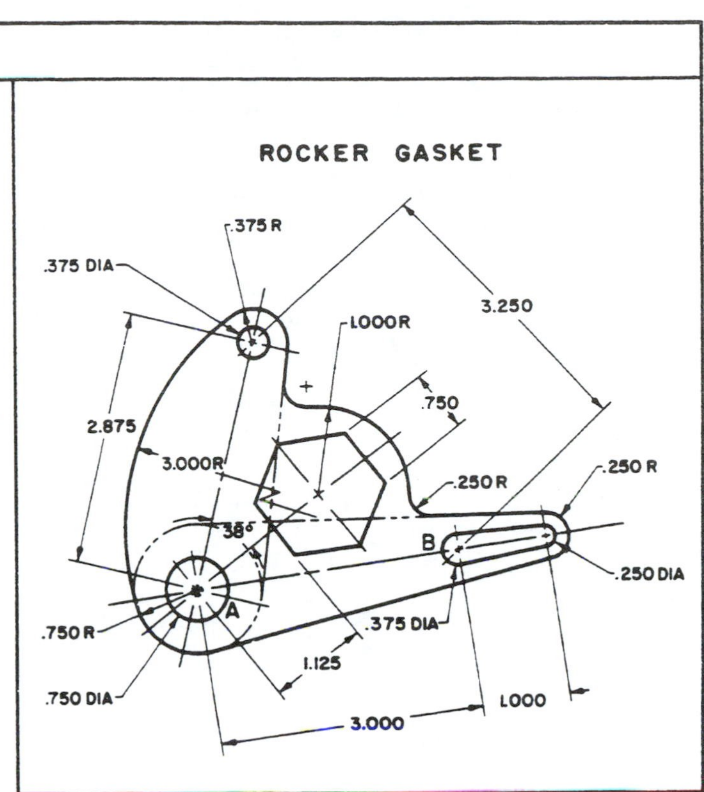

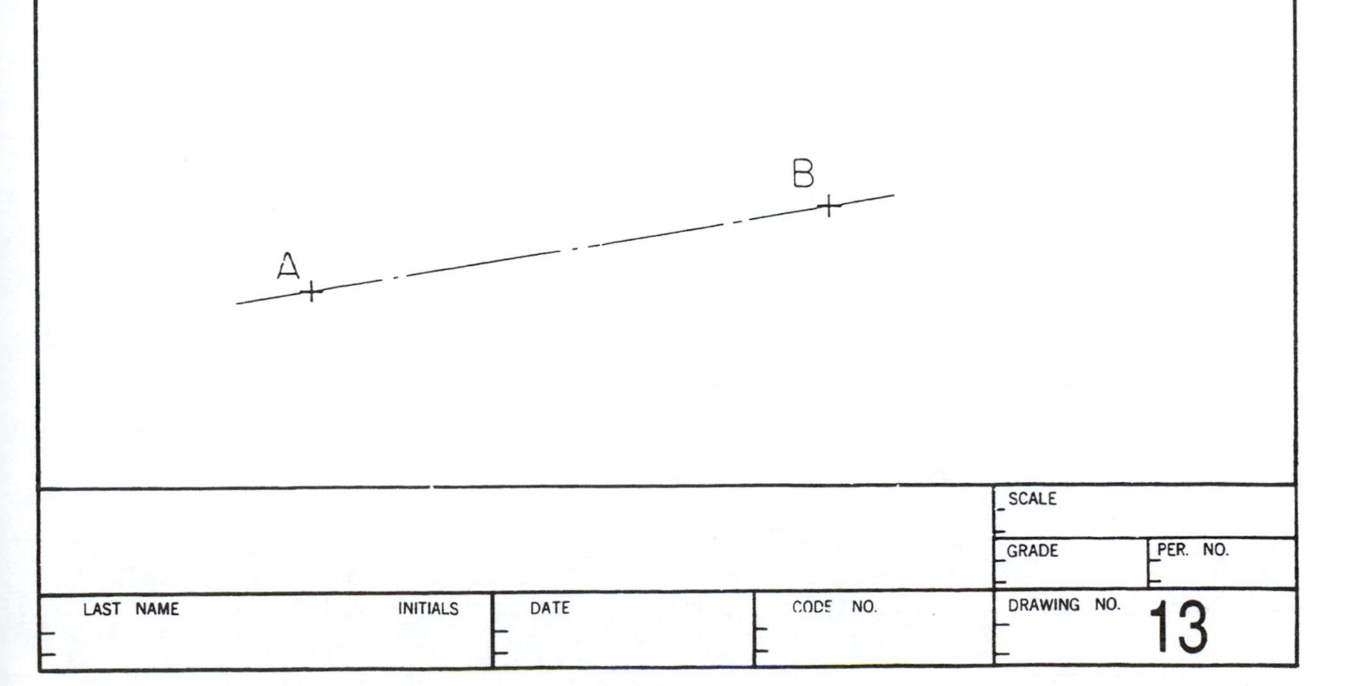

SCALE	
GRADE	PER. NO.
DRAWING NO.	13

LAST NAME	INITIALS	DATE	CODE NO.

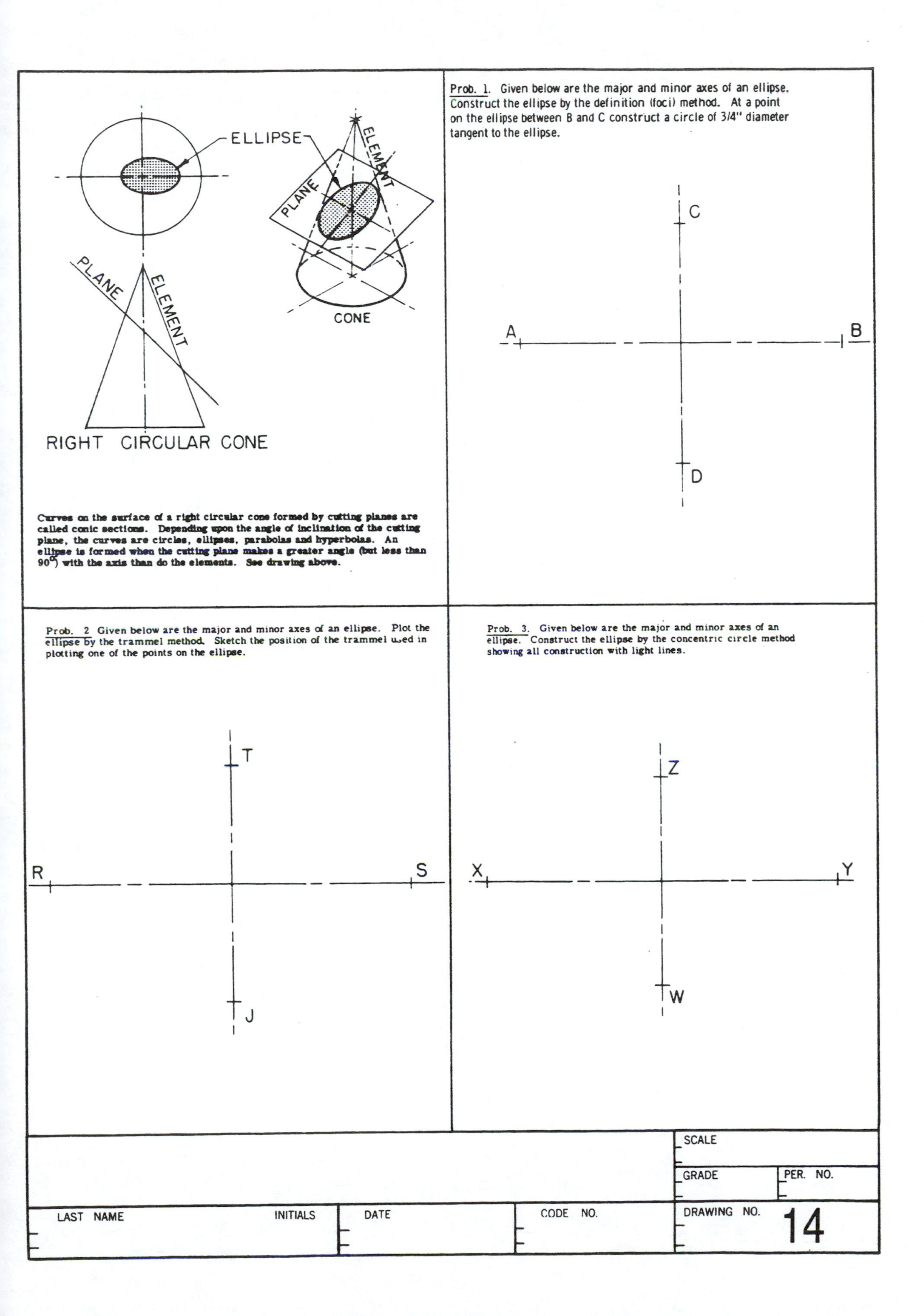

Curves on the surface of a right circular cone formed by cutting planes are called conic sections. Depending upon the angle of inclination of the cutting plane, the curves are circles, ellipses, parabolas and hyperbolas. An ellipse is formed when the cutting plane makes a greater angle (but less than 90°) with the axis than do the elements. See drawing above.

Prob. 1. Given below are the major and minor axes of an ellipse. Construct the ellipse by the definition (foci) method. At a point on the ellipse between B and C construct a circle of 3/4" diameter tangent to the ellipse.

Prob. 2 Given below are the major and minor axes of an ellipse. Plot the ellipse by the trammel method. Sketch the position of the trammel used in plotting one of the points on the ellipse.

Prob. 3. Given below are the major and minor axes of an ellipse. Construct the ellipse by the concentric circle method showing all construction with light lines.

LAST NAME	INITIALS	DATE	CODE NO.	SCALE / GRADE / PER. NO. / DRAWING NO. 14

Given below is a one view drawing of a "Linkage" composed of four parts. The point of linkage slides back and forth in a horizontal groove as the arm revolves clockwise and counterclockwise about the center A. The movement of the arm may extend counterclockwise and clockwise to contact the solid rods C and D respectively.

a) Draw the arm revolved in a counterclockwise direction until it comes in contact with rod C.
b) Draw the arm revolved in a clockwise direction until it comes in contact with rod D.
c) Determine and record the total distance of travel of the linkage for the extreme contact points of the arm with rod C and rod D. ______
d) Measure and record the maximum angle the arm revolves from contact to contact. ______

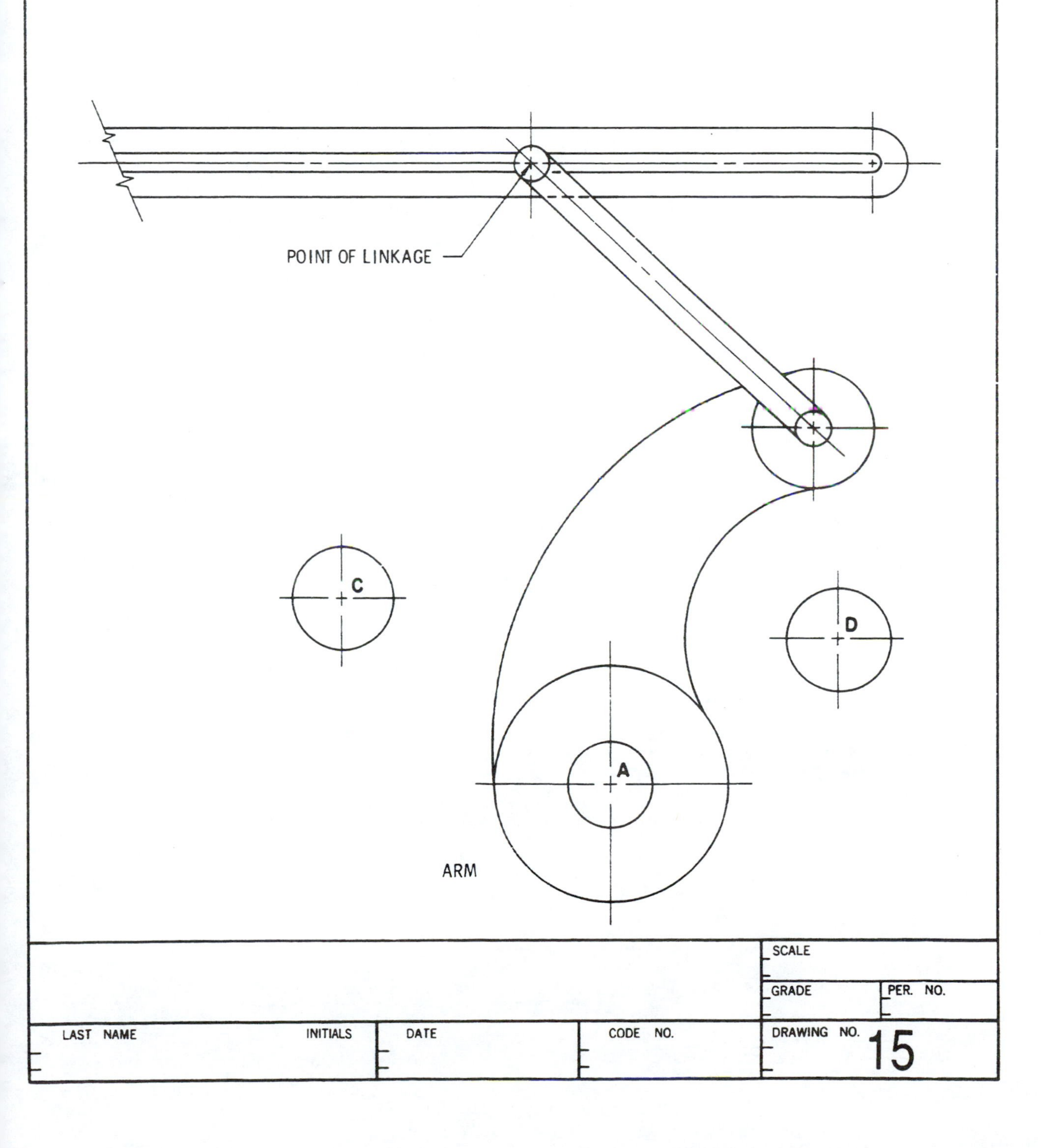

	SCALE	
	GRADE	PER. NO.
LAST NAME INITIALS	DATE	CODE NO.	DRAWING NO. 15

PICTORIAL VIEWS

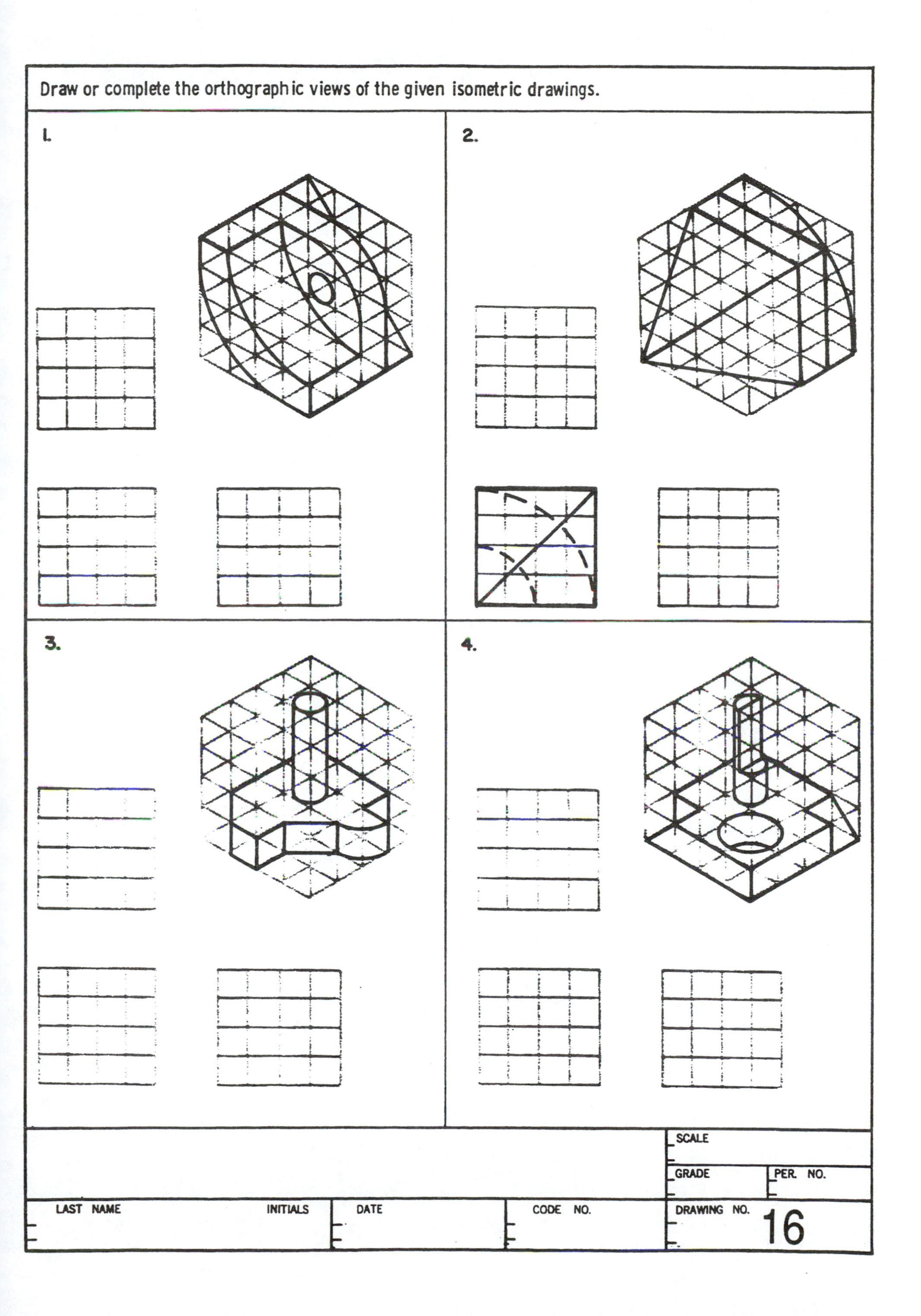
Draw or complete the orthographic views of the given isometric drawings.
1.
2.
3.
4.
SCALE
GRADE
PER. NO.
LAST NAME
INITIALS
DATE
CODE NO.
DRAWING NO.
16

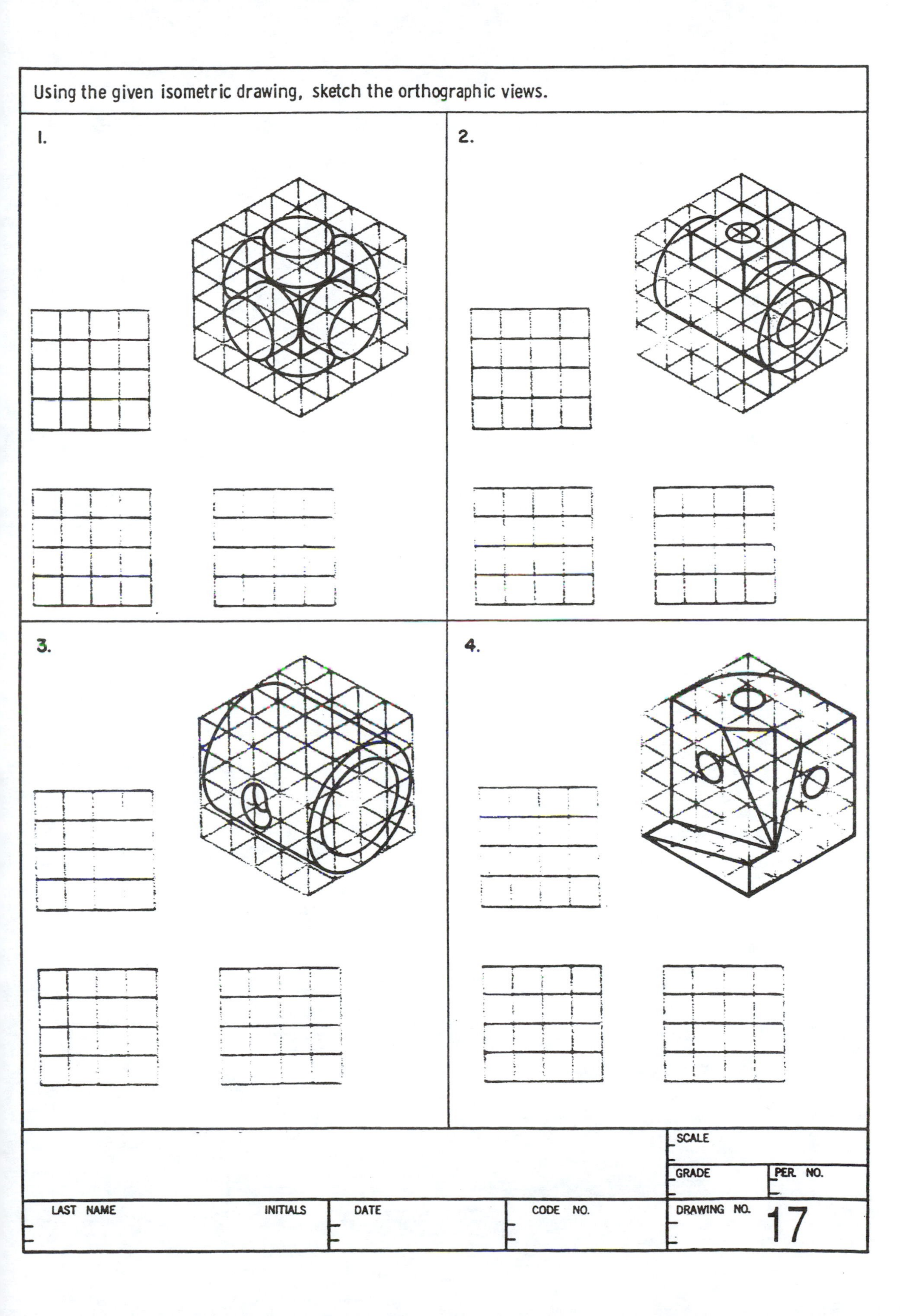

Using the given isometric drawing, sketch the orthographic views.
1.
2.
3.
4.
SCALE
GRADE
PER. NO.
LAST NAME
INITIALS
DATE
CODE NO.
DRAWING NO.
17

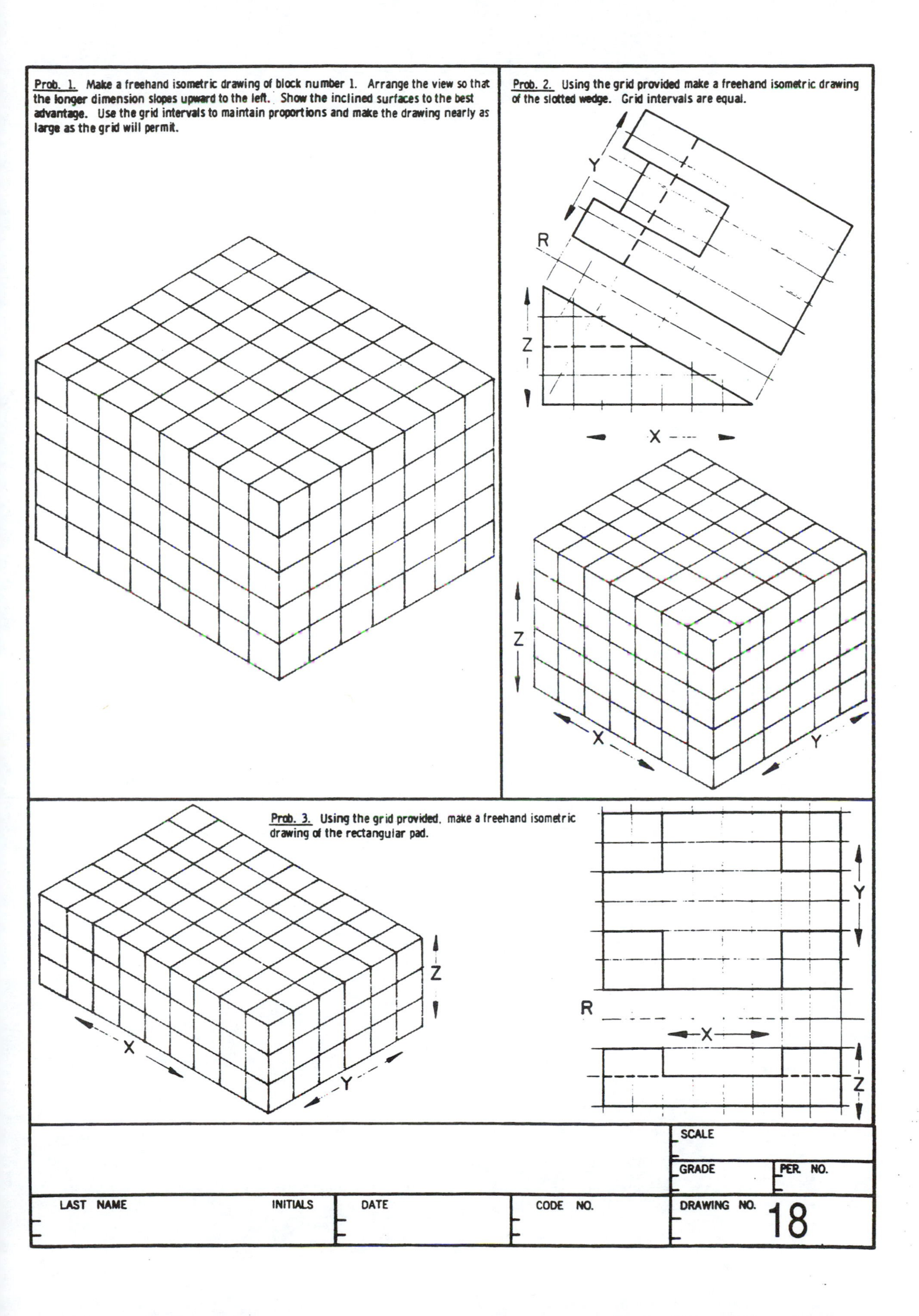
Prob. 1. Make a freehand isometric drawing of block number 1. Arrange the view so that the longer dimension slopes upward to the left. Show the inclined surfaces to the best advantage. Use the grid intervals to maintain proportions and make the drawing nearly as large as the grid will permit.
Prob. 2. Using the grid provided make a freehand isometric drawing of the slotted wedge. Grid intervals are equal.
Y
R
Z
X
Z
X
Y
Prob. 3. Using the grid provided, make a freehand isometric drawing of the rectangular pad.
Z
X
Y
Y
R
X
Z
SCALE
GRADE
PER. NO.
LAST NAME
INITIALS
DATE
CODE NO.
DRAWING NO.
18

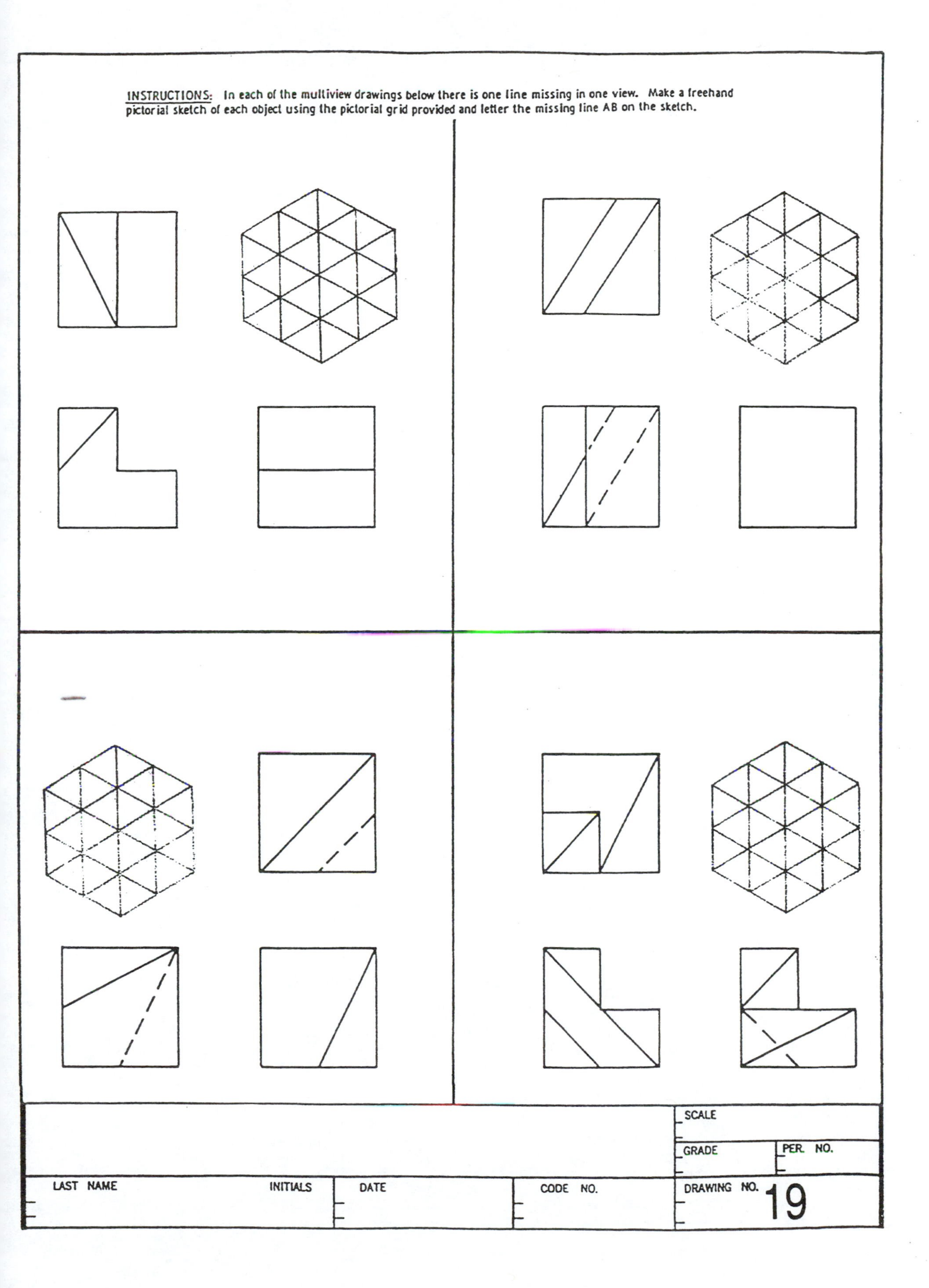

INSTRUCTIONS: In each of the multiview drawings below there is one line missing in one view. Make a freehand pictorial sketch of each object using the pictorial grid provided and letter the missing line AB on the sketch.

			SCALE	
			GRADE	PER. NO.
LAST NAME INITIALS	DATE	CODE NO.	DRAWING NO. 19	

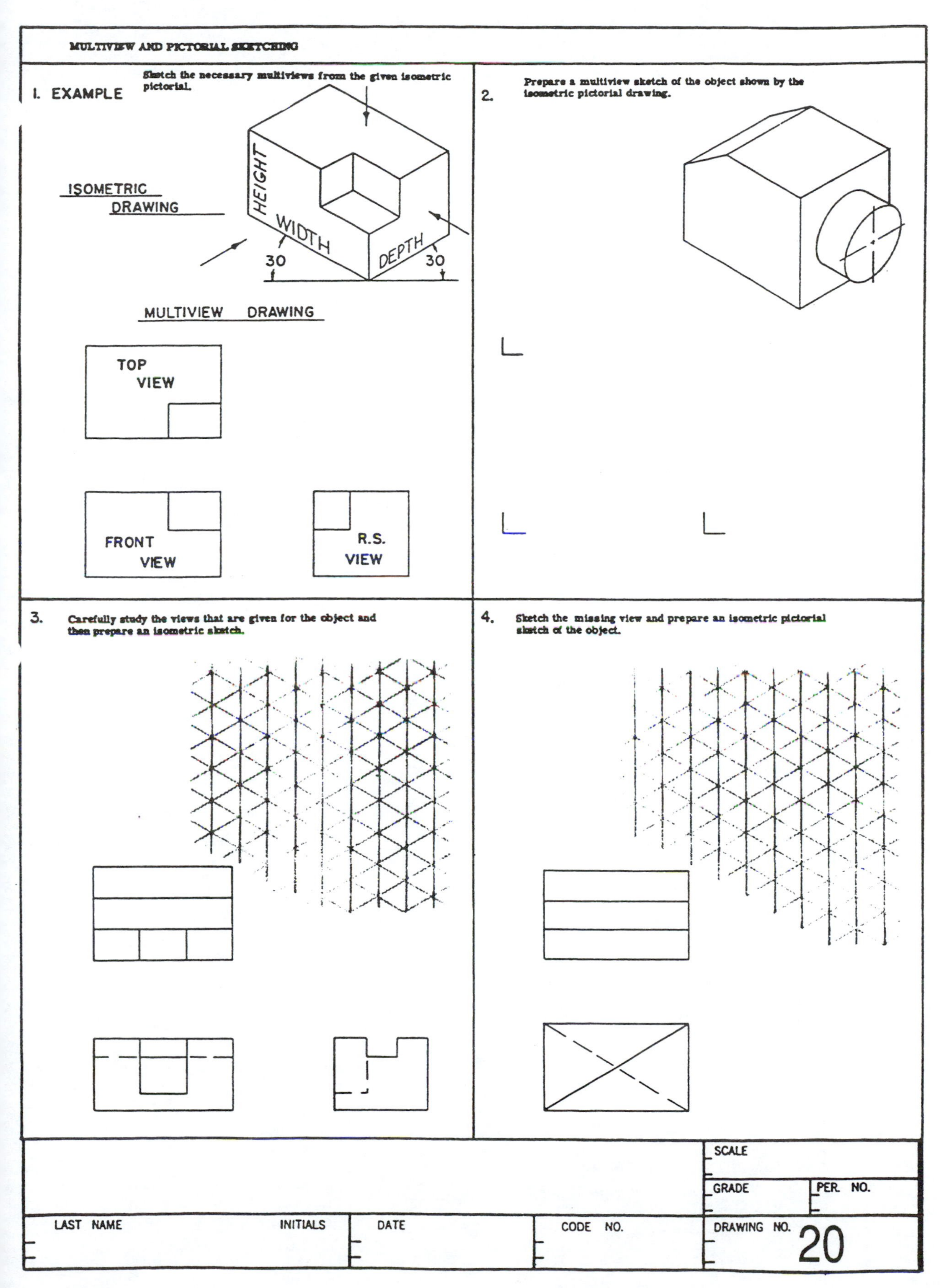
MULTIVIEW AND PICTORIAL SKETCHING
1. EXAMPLE
Sketch the necessary multiviews from the given isometric pictorial.
ISOMETRIC DRAWING
HEIGHT
WIDTH
DEPTH
30
30
MULTIVIEW DRAWING
TOP VIEW
FRONT VIEW
R.S. VIEW
2.
Prepare a multiview sketch of the object shown by the isometric pictorial drawing.
3.
Carefully study the views that are given for the object and then prepare an isometric sketch.
4.
Sketch the missing view and prepare an isometric pictorial sketch of the object.
SCALE
GRADE
PER. NO.
LAST NAME
INITIALS
DATE
CODE NO.
DRAWING NO.
20

Make an instrument drawing of the third view for each problem. The views, as given, are correct and complete. Make a freehand isometric sketch of the object as you visualize it.

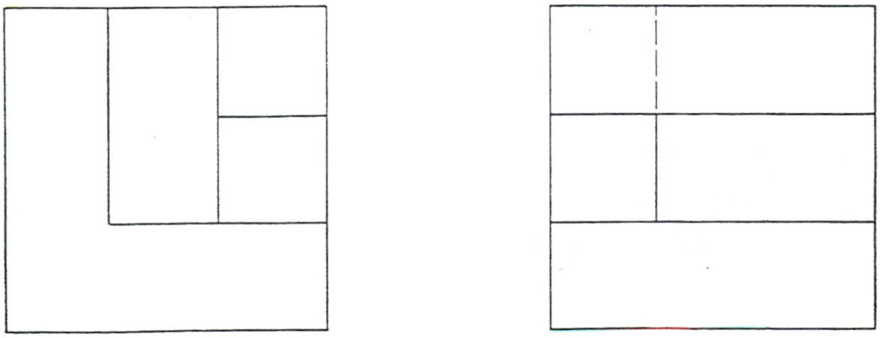

SCALE		
GRADE	PER. NO.	

LAST NAME	INITIALS	DATE	CODE NO.	DRAWING NO. 21

INSTRUCTIONS:

(1) Make a freehand isometric study sketch of the given object in the grid area. Use reasonable care in the execution and proportioning of the sketch.

(2) Using instruments, draw the top view and letter all views of the points of the right hand notch.

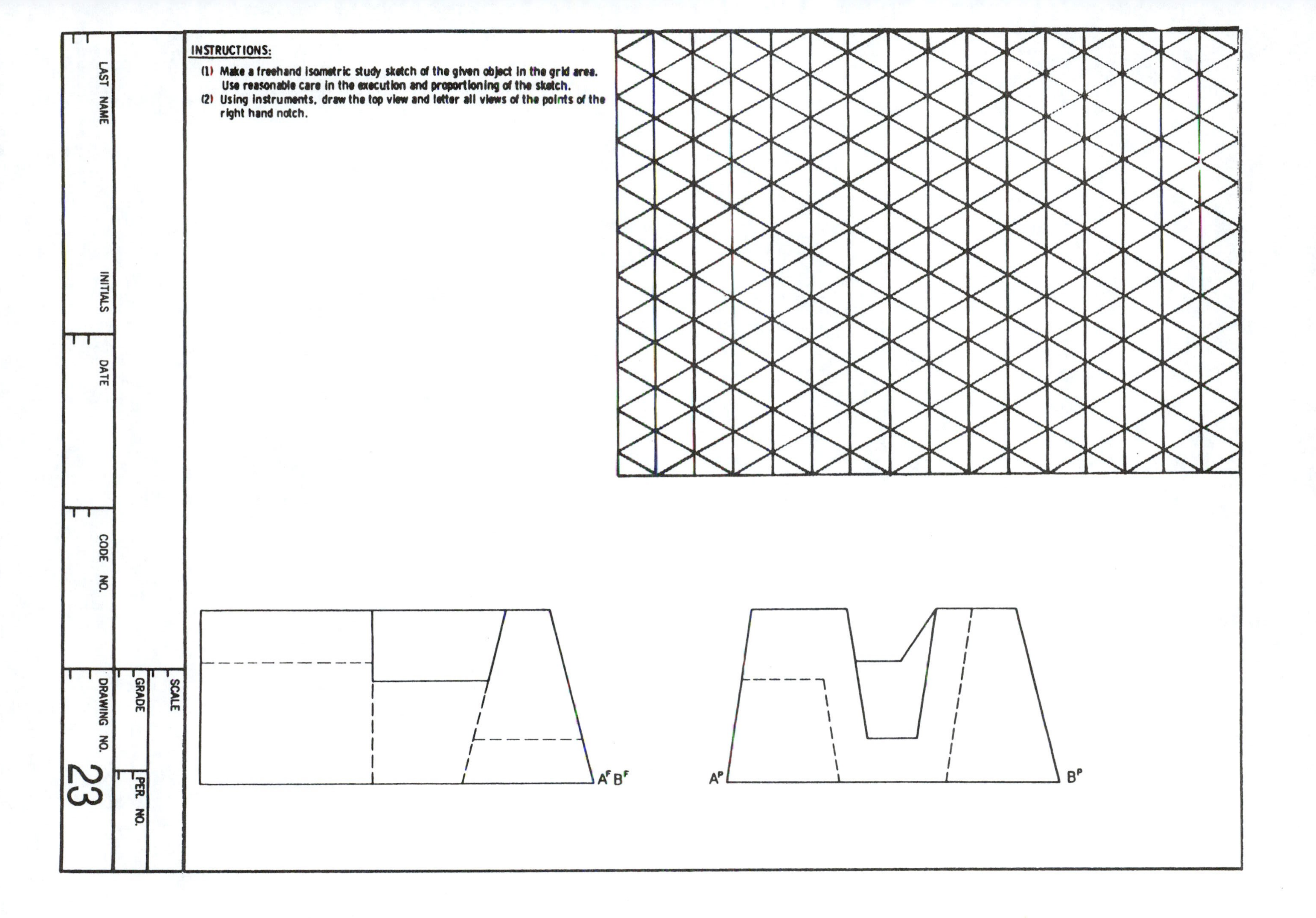

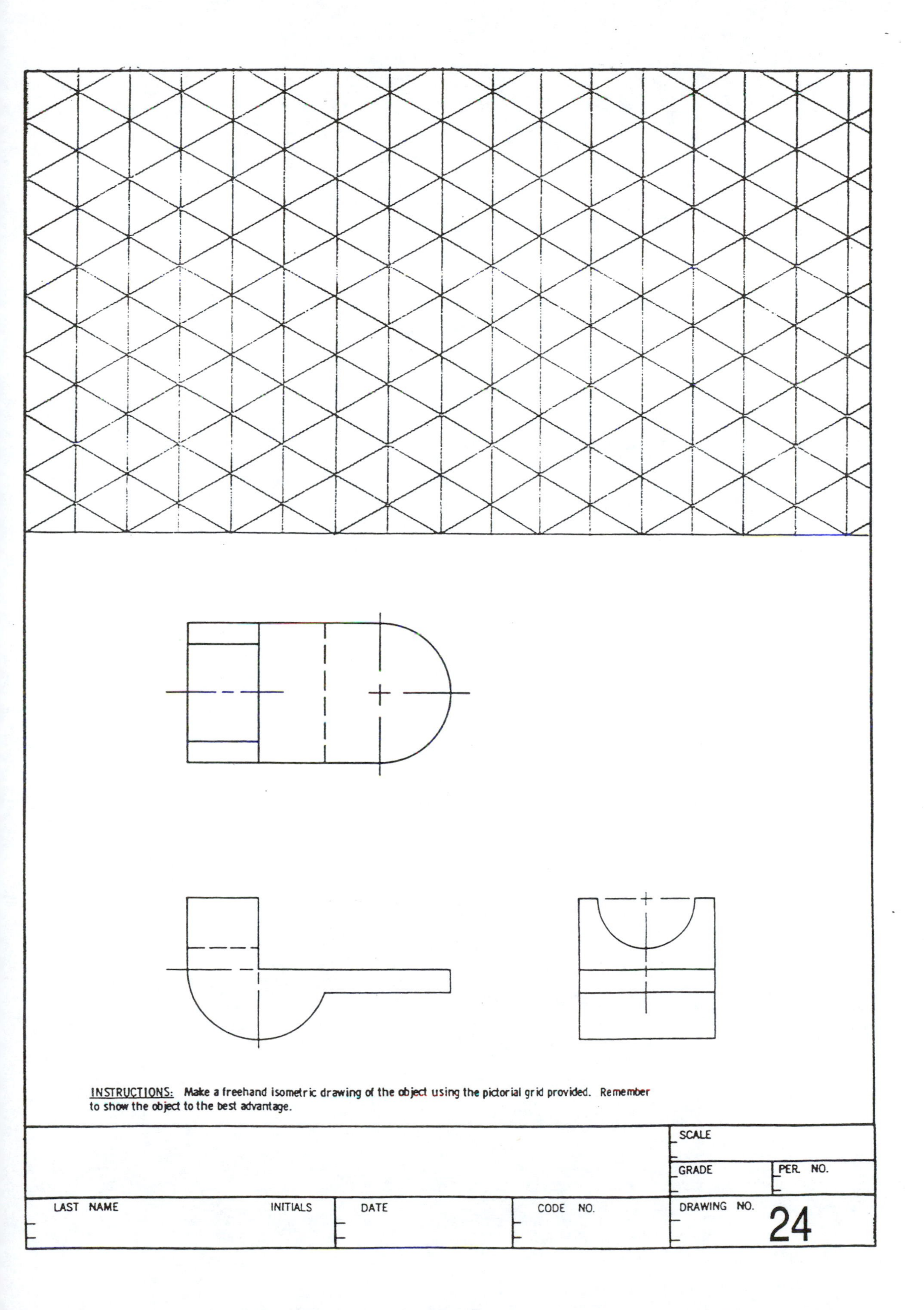
INSTRUCTIONS: Make a freehand isometric drawing of the object using the pictorial grid provided. Remember to show the object to the best advantage.
SCALE
GRADE
PER. NO.
LAST NAME
INITIALS
DATE
CODE NO.
DRAWING NO.
24

Utilizing the information provided by the drawing of the Bracket, execute an isometric drawing of the object on an "A" size paper. It should be noted that the object is composed mostly of circles, arcs and irregular curves. While executing the drawing keep in mind the following:

a) Select an isometric position which will show the greatest information about the object.

b) Construct the 1" R arc by the four-center method. Construct all holes through the use of the isometric template. Construct the irregular curve by the offset method. Construct the 2" Dia. circle by the four-center method.

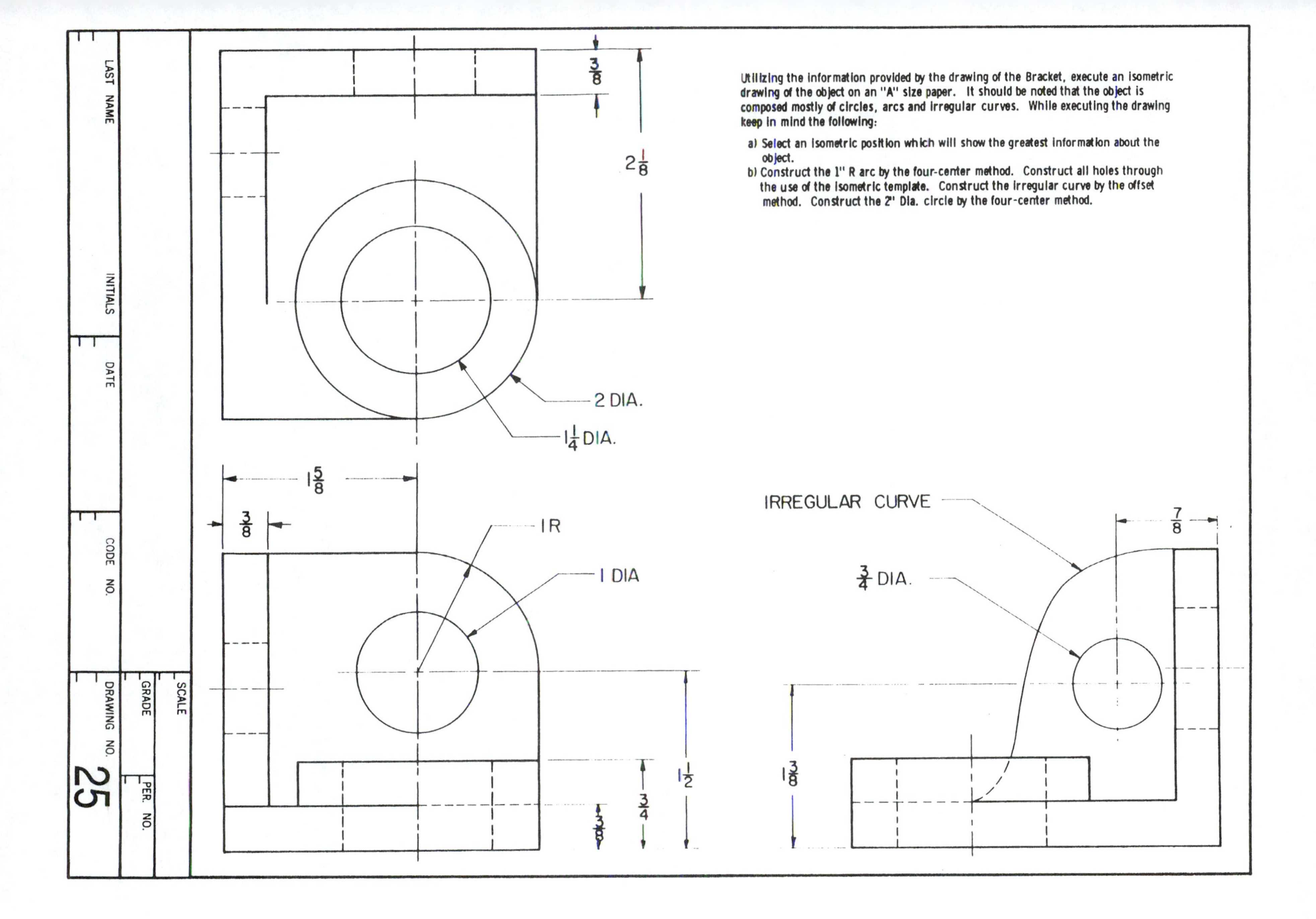

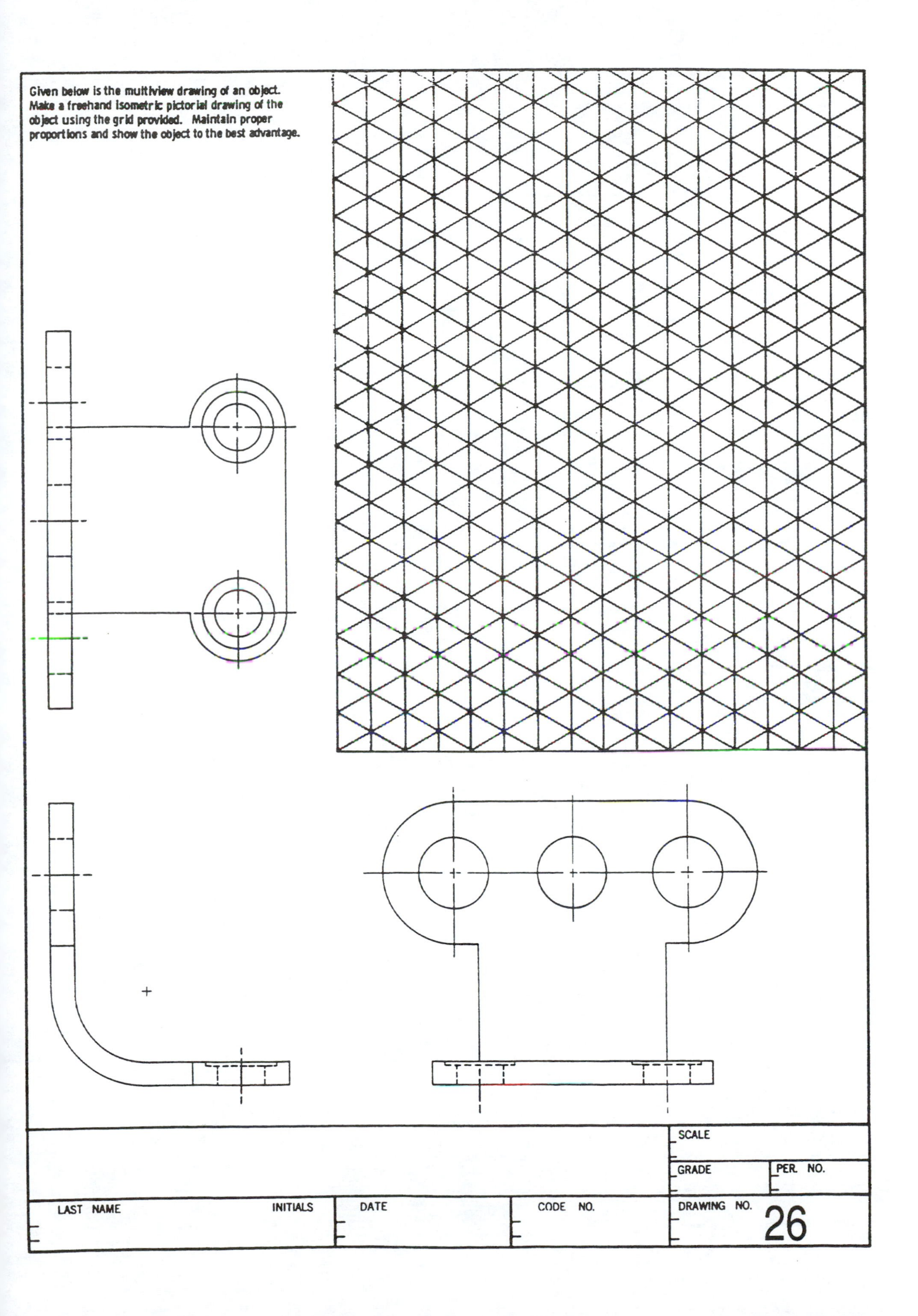
Given below is the multiview drawing of an object. Make a freehand isometric pictorial drawing of the object using the grid provided. Maintain proper proportions and show the object to the best advantage.
SCALE
GRADE
PER. NO.
LAST NAME
INITIALS
DATE
CODE NO.
DRAWING NO.
26

MULTIVIEWS

Make freehand multiview drawings which show the complete shape description of the given objects. Start the drawings at the indicated positions below their respective pictorials. The scale intervals along the sides of the pictorials are given primarily for proportions rather than numerical scale values.

A freehand drawing is not made to any particular scale except when it is made on a grid of scaleable units.

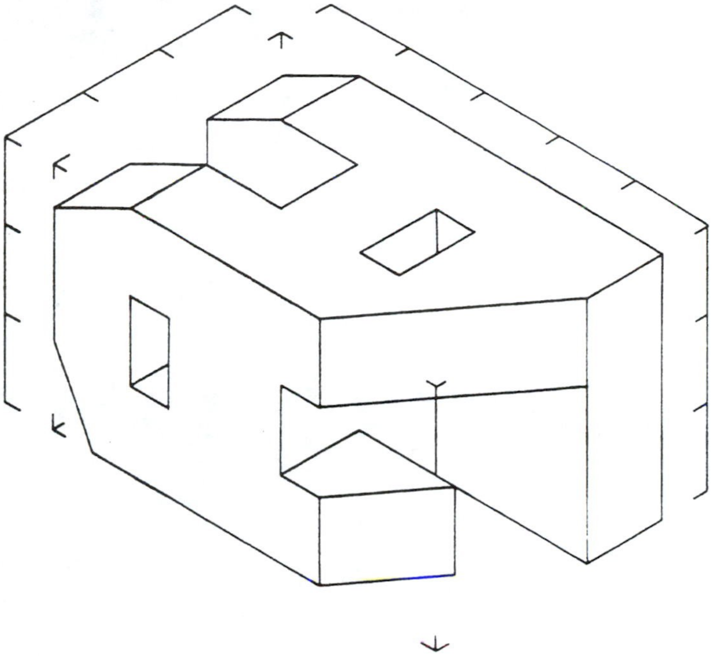

LAST NAME	INITIALS	DATE	CODE NO.	SCALE	GRADE	PER. NO.	DRAWING NO.
							28

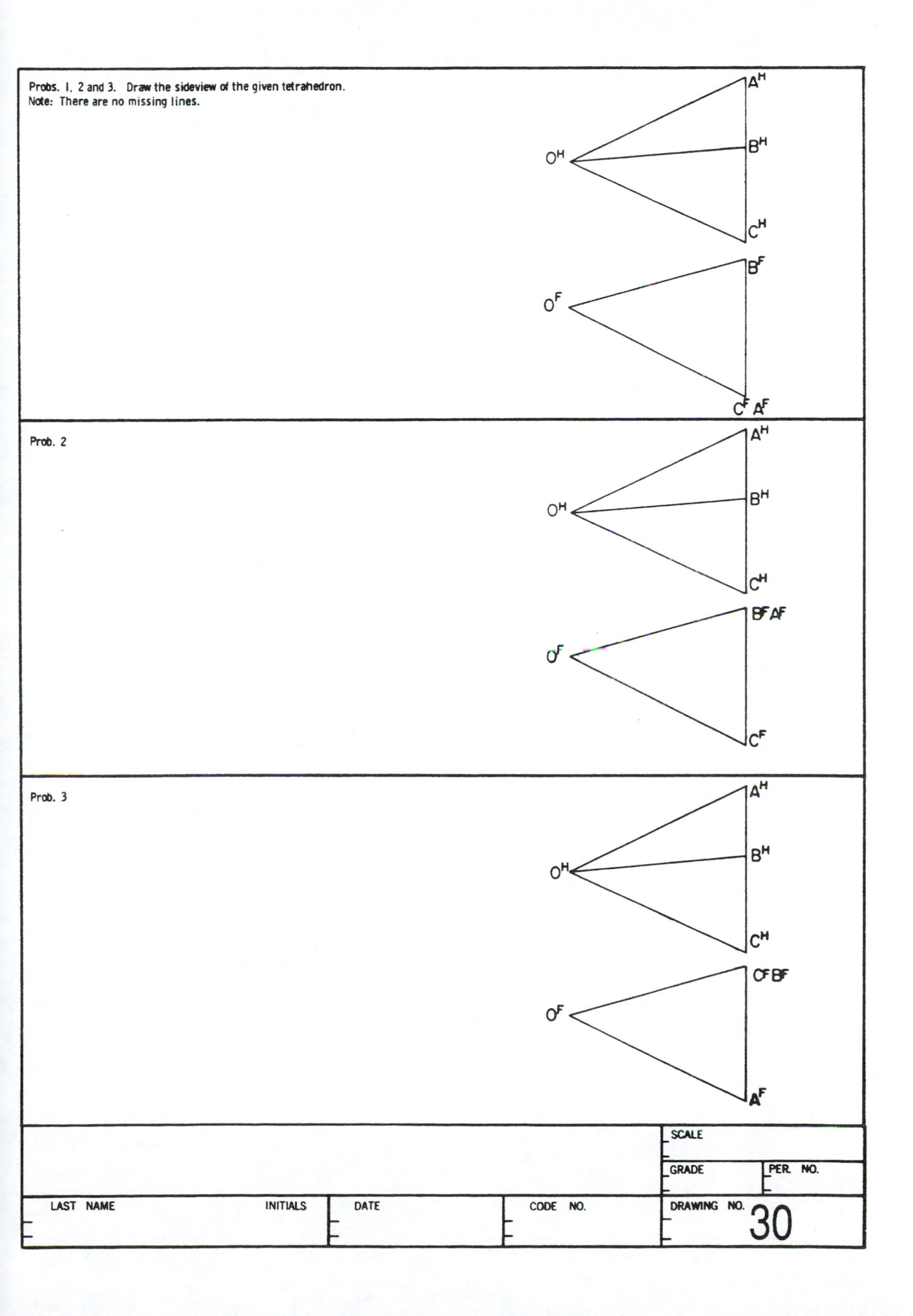
Probs. 1, 2 and 3. Draw the sideview of the given tetrahedron.
Note: There are no missing lines.
A^H
B^H
O^H
C^H
B^F
O^F
C^F A^F
Prob. 2
A^H
B^H
O^H
C^H
$B^F A^F$
O^F
C^F
Prob. 3
A^H
B^H
O^H
C^H
$C^F B^F$
O^F
A^F
SCALE
GRADE
PER. NO.
LAST NAME
INITIALS
DATE
CODE NO.
DRAWING NO.
30

Prob. 1. Given the left end view of the irregular block. Draw the front and top views using the dimensions shown on the out-of-size pictorial. Use Scale 1" = 2".

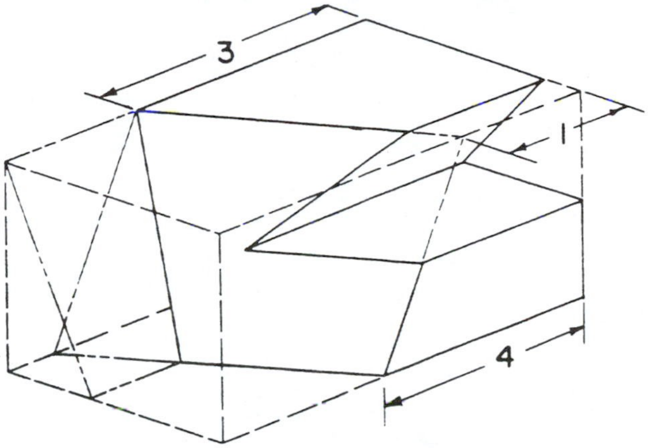

3

1

4

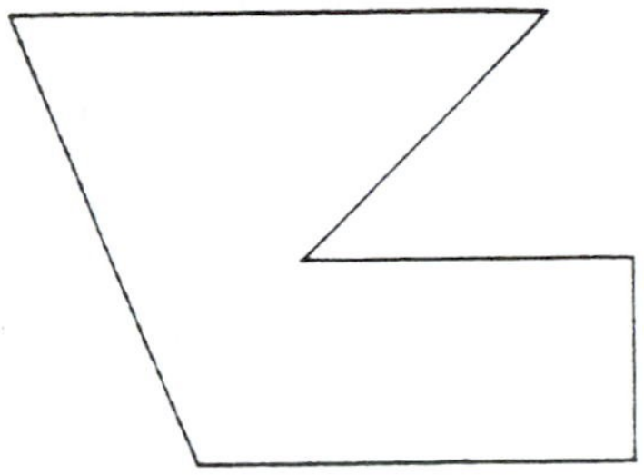

Prob. 2. Make a full size multiview drawing of the left hand portion of the object after it has been cut by a plane defined by points A, B, and C. Note: The pictorial is not to scale and the parallel ends on the object lie in the ends of the enclosing parallelepiped. Point A has been located in three views.

A^H

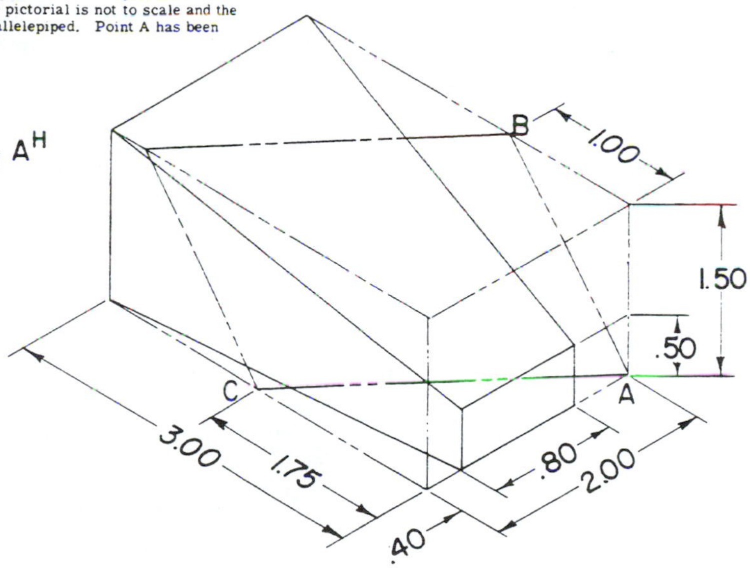

B

1.00

1.50

.50

A

C

3.00

1.75

.80

2.00

.40

A^F

A^P

	SCALE	
	GRADE	PER. NO.

LAST NAME	INITIALS	DATE	CODE NO.	DRAWING NO. 31

INSTRUCTIONS: Complete the pictorial sketch by adding the missing lines of intersection between the surfaces of the through rectangular hole and the sloping surface. Make a three view multiview drawing of the Control Block.
Scale: 1" = 2"

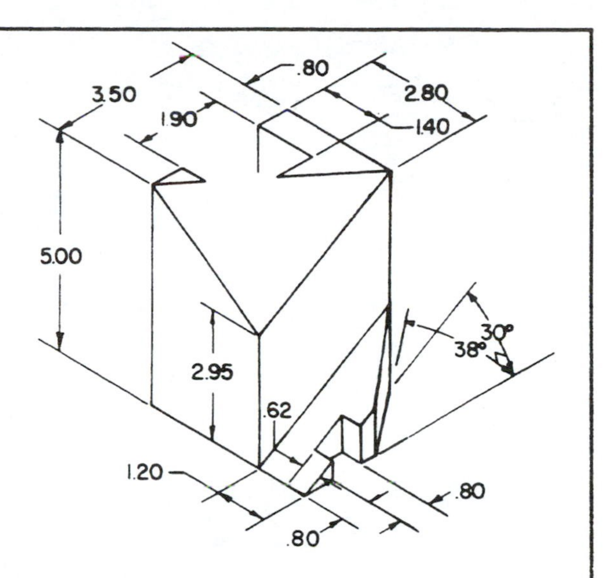

FORM SA

	SCALE	
	GRADE	PER. NO.

LAST NAME	INITIALS	DATE	CODE NO.	DRAWING NO. 32

INSTRUCTIONS: Make a freehand isometric study sketch.
Then, with instruments, draw the missing view.

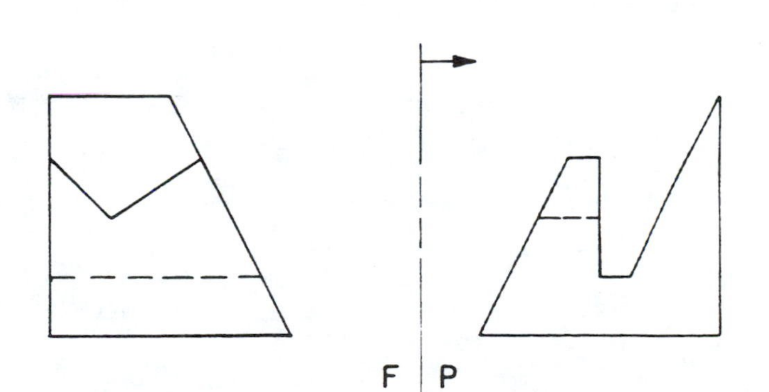

LAST NAME INITIALS	DATE	CODE NO.	SCALE	
			GRADE	PER. NO.
			DRAWING NO. 33	

AUXILIARY VIEWS

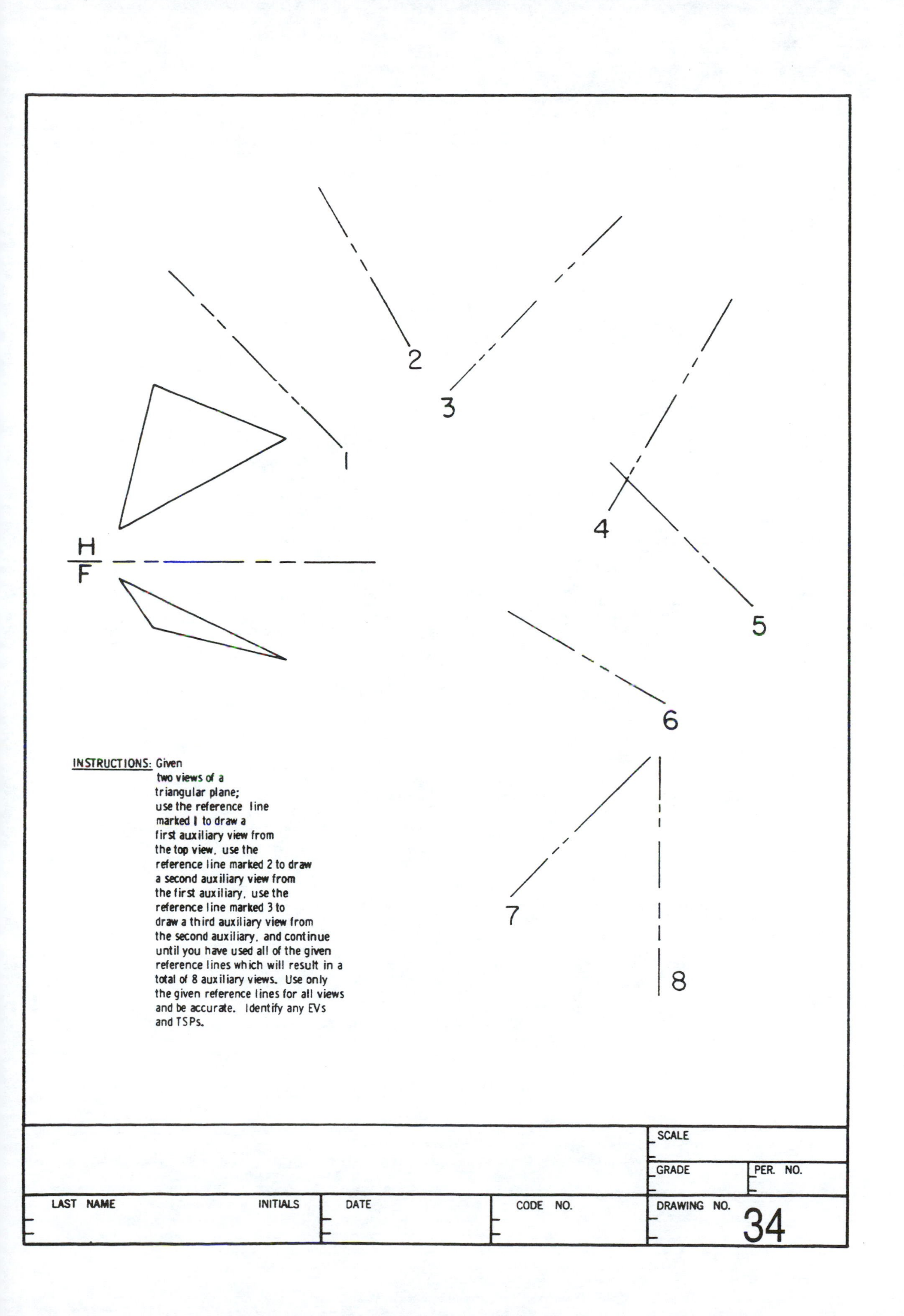
2
3
1
4
5
H
F
6
7
8
INSTRUCTIONS: Given two views of a triangular plane; use the reference line marked 1 to draw a first auxiliary view from the top view, use the reference line marked 2 to draw a second auxiliary view from the first auxiliary, use the reference line marked 3 to draw a third auxiliary view from the second auxiliary, and continue until you have used all of the given reference lines which will result in a total of 8 auxiliary views. Use only the given reference lines for all views and be accurate. Identify any EVs and TSPs.
SCALE
GRADE
PER. NO.
LAST NAME
INITIALS
DATE
CODE NO.
DRAWING NO.
34

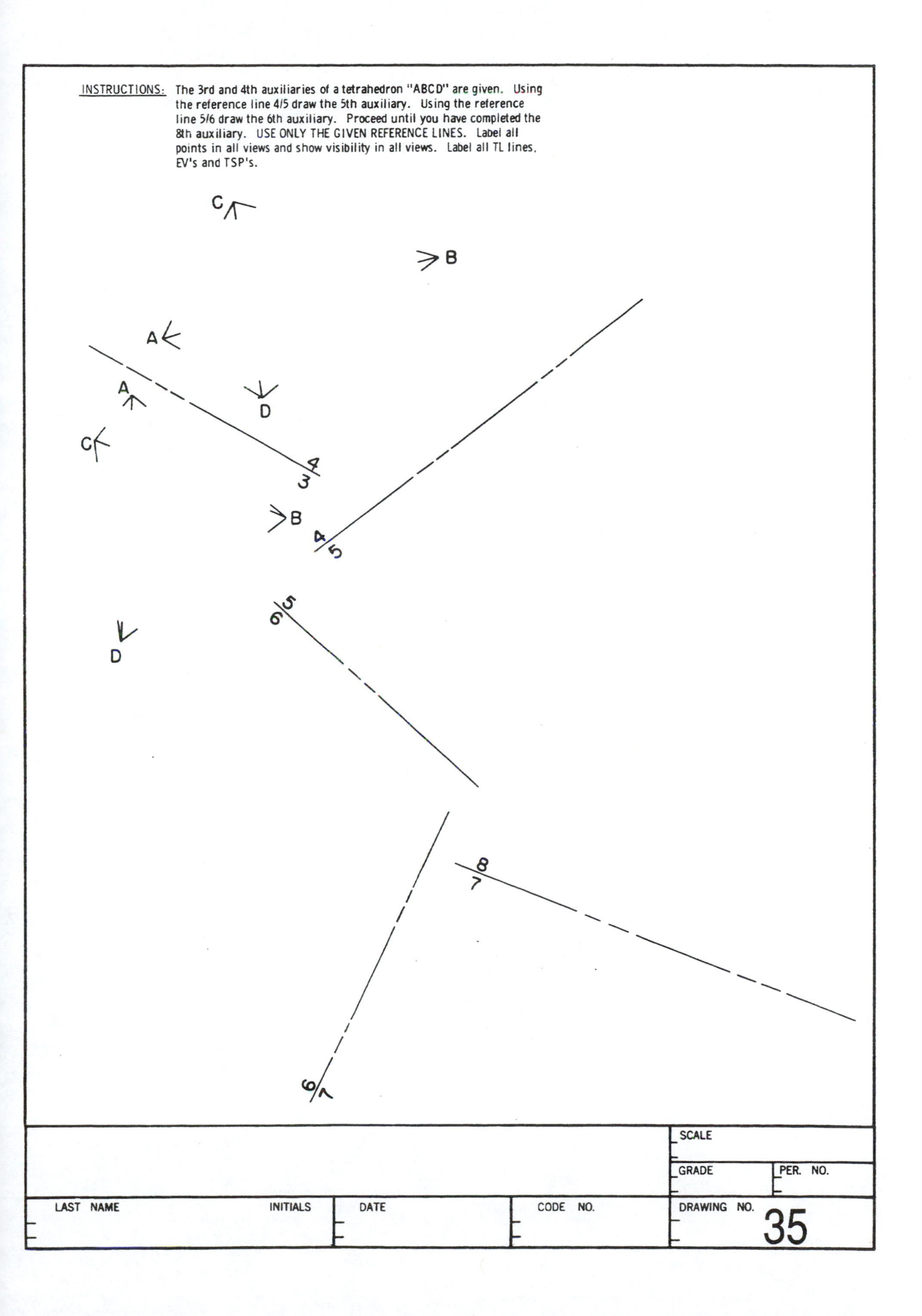
INSTRUCTIONS: The 3rd and 4th auxiliaries of a tetrahedron "ABCD" are given. Using the reference line 4/5 draw the 5th auxiliary. Using the reference line 5/6 draw the 6th auxiliary. Proceed until you have completed the 8th auxiliary. USE ONLY THE GIVEN REFERENCE LINES. Label all points in all views and show visibility in all views. Label all TL lines, EV's and TSP's.
C
B
A
A
D
C
4
3
B
4
5
5
6
D
8
7
6
7
SCALE
GRADE
PER. NO.
LAST NAME
INITIALS
DATE
CODE NO.
DRAWING NO.
35

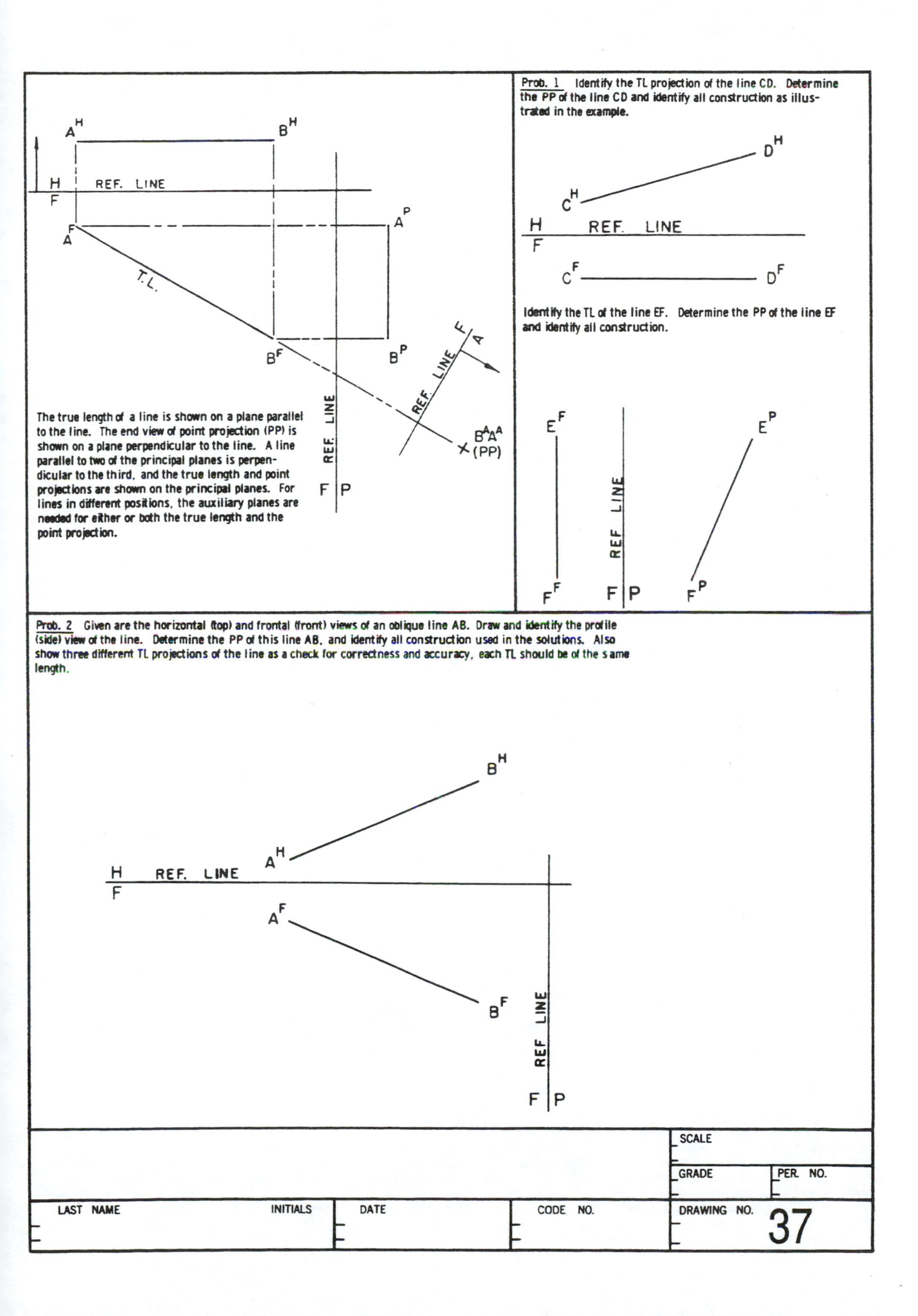

The true length of a line is shown on a plane parallel to the line. The end view of point projection (PP) is shown on a plane perpendicular to the line. A line parallel to two of the principal planes is perpendicular to the third, and the true length and point projections are shown on the principal planes. For lines in different positions, the auxiliary planes are needed for either or both the true length and the point projection.

Prob. 1 Identify the TL projection of the line CD. Determine the PP of the line CD and identify all construction as illustrated in the example.

Identify the TL of the line EF. Determine the PP of the line EF and identify all construction.

Prob. 2 Given are the horizontal (top) and frontal (front) views of an oblique line AB. Draw and identify the profile (side) view of the line. Determine the PP of this line AB, and identify all construction used in the solutions. Also show three different TL projections of the line as a check for correctness and accuracy, each TL should be of the same length.

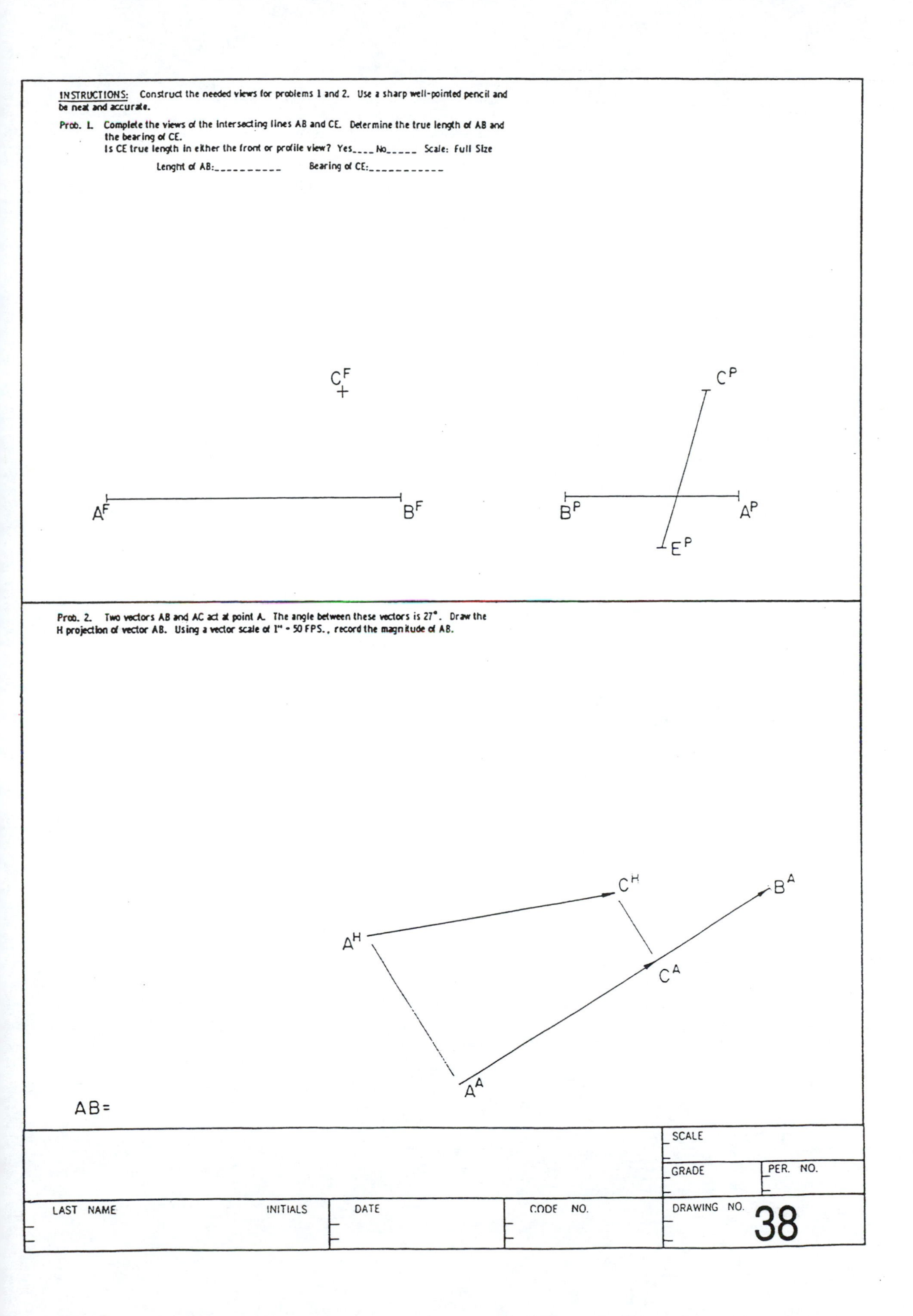

INSTRUCTIONS: Construct the needed views for problems 1 and 2. Use a sharp well-pointed pencil and be neat and accurate.

Prob. 1. Complete the views of the intersecting lines AB and CE. Determine the true length of AB and the bearing of CE.
Is CE true length in either the front or profile view? Yes____ No_____ Scale: Full Size

Lenght of AB:__________ Bearing of CE:___________

Prob. 2. Two vectors AB and AC act at point A. The angle between these vectors is 27°. Draw the H projection of vector AB. Using a vector scale of 1" = 50 FPS., record the magnitude of AB.

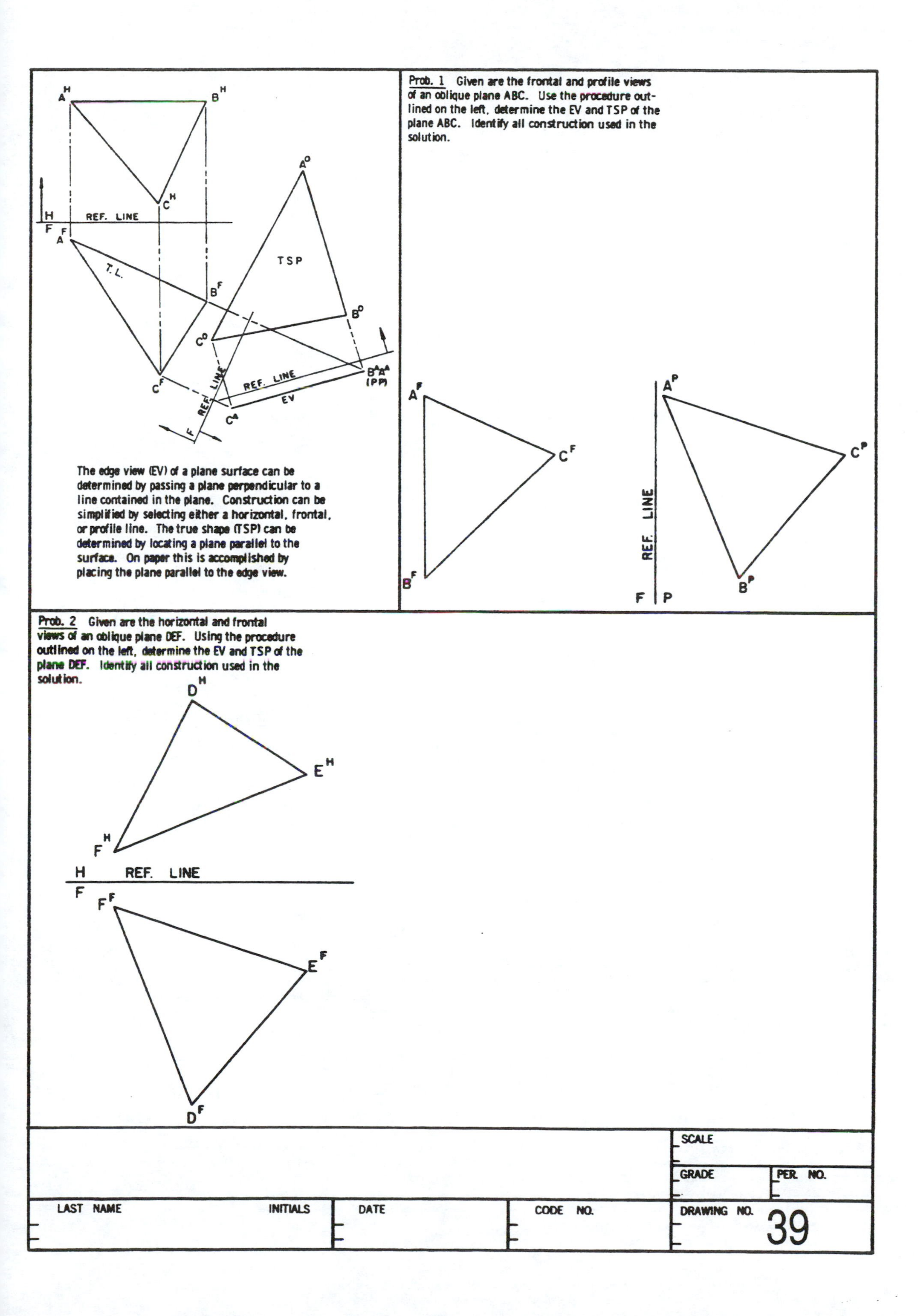

A^H
B^H
C^H
H
REF. LINE
F
A^F
T.L.
B^F
C^F
A^O
TSP
B^O
C^O
REF. LINE
B^A A^A
(PP)
EV
REF. LINE
C^A
F
The edge view (EV) of a plane surface can be determined by passing a plane perpendicular to a line contained in the plane. Construction can be simplified by selecting either a horizontal, frontal, or profile line. The true shape (TSP) can be determined by locating a plane parallel to the surface. On paper this is accomplished by placing the plane parallel to the edge view.
Prob. 1 Given are the frontal and profile views of an oblique plane ABC. Use the procedure outlined on the left, determine the EV and TSP of the plane ABC. Identify all construction used in the solution.
A^F
C^F
B^F
A^P
C^P
B^P
REF. LINE
F
P
Prob. 2 Given are the horizontal and frontal views of an oblique plane DEF. Using the procedure outlined on the left, determine the EV and TSP of the plane DEF. Identify all construction used in the solution.
D^H
E^H
F^H
H
REF. LINE
F
F^F
E^F
D^F
SCALE
GRADE
PER. NO.
LAST NAME
INITIALS
DATE
CODE NO.
DRAWING NO.
39

Determine whether point "D" lies in the plane "ABC." If it does, record its distance from side "AC" where measurable. If it does not, record its distance from the plane where measurable.

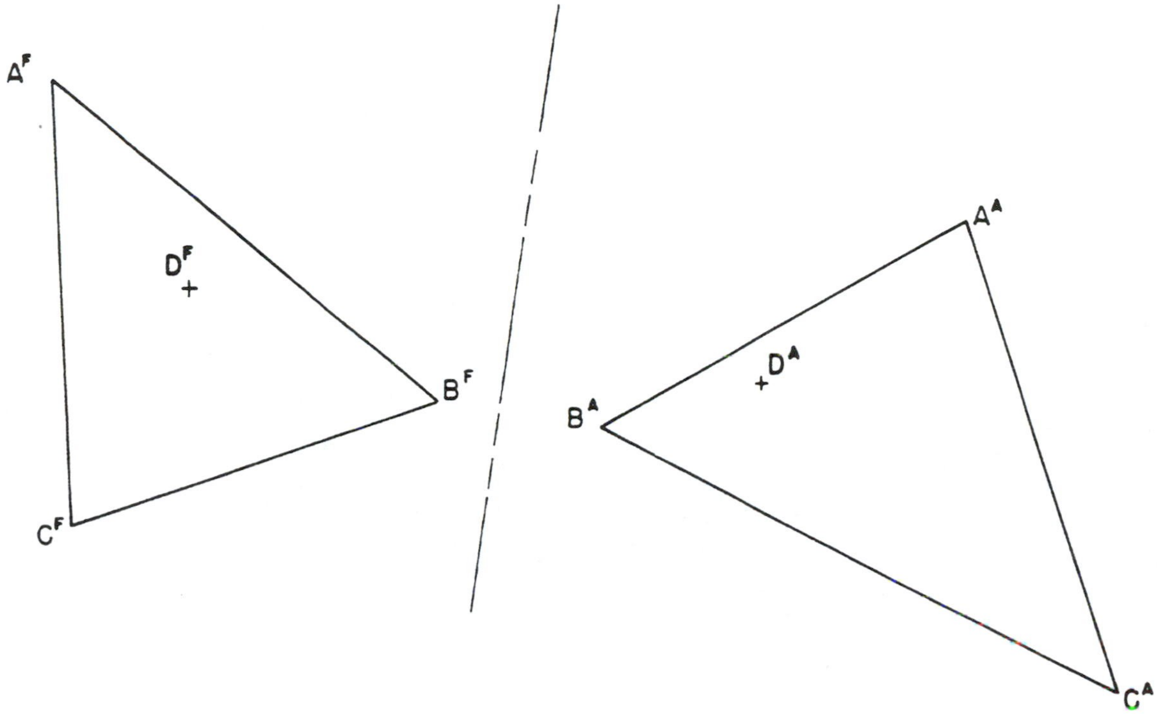

A^F

D^F

B^F

C^F

A^A

D^A

B^A

C^A

	SCALE	
	GRADE	PER. NO.

LAST NAME INITIALS	DATE	CODE NO.	DRAWING NO. 40

SCALE: 1" = 1"

INSTRUCTIONS: Complete the front and top views of parallelogram "ABCD." Diagonal "AB" is 3.2" long. Determine and record where measurable, the true length of diagonal "CD."

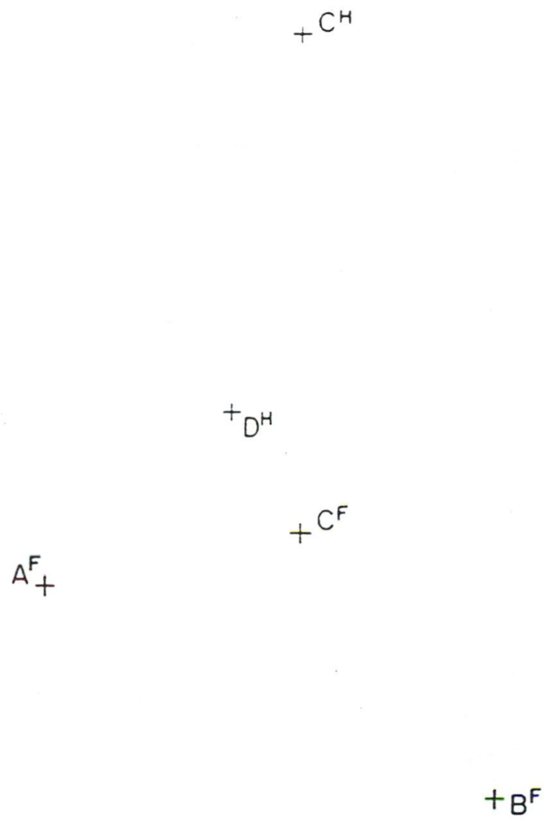

LAST NAME INITIALS	DATE	CODE NO.	SCALE GRADE / PER. NO. DRAWING NO. 42

SCALE: Full Size

INSTRUCTIONS: In each problem below draw the view which will show the true shape of the inclined surface. Do not show the entire object. Indicate the reference lines used as RL.

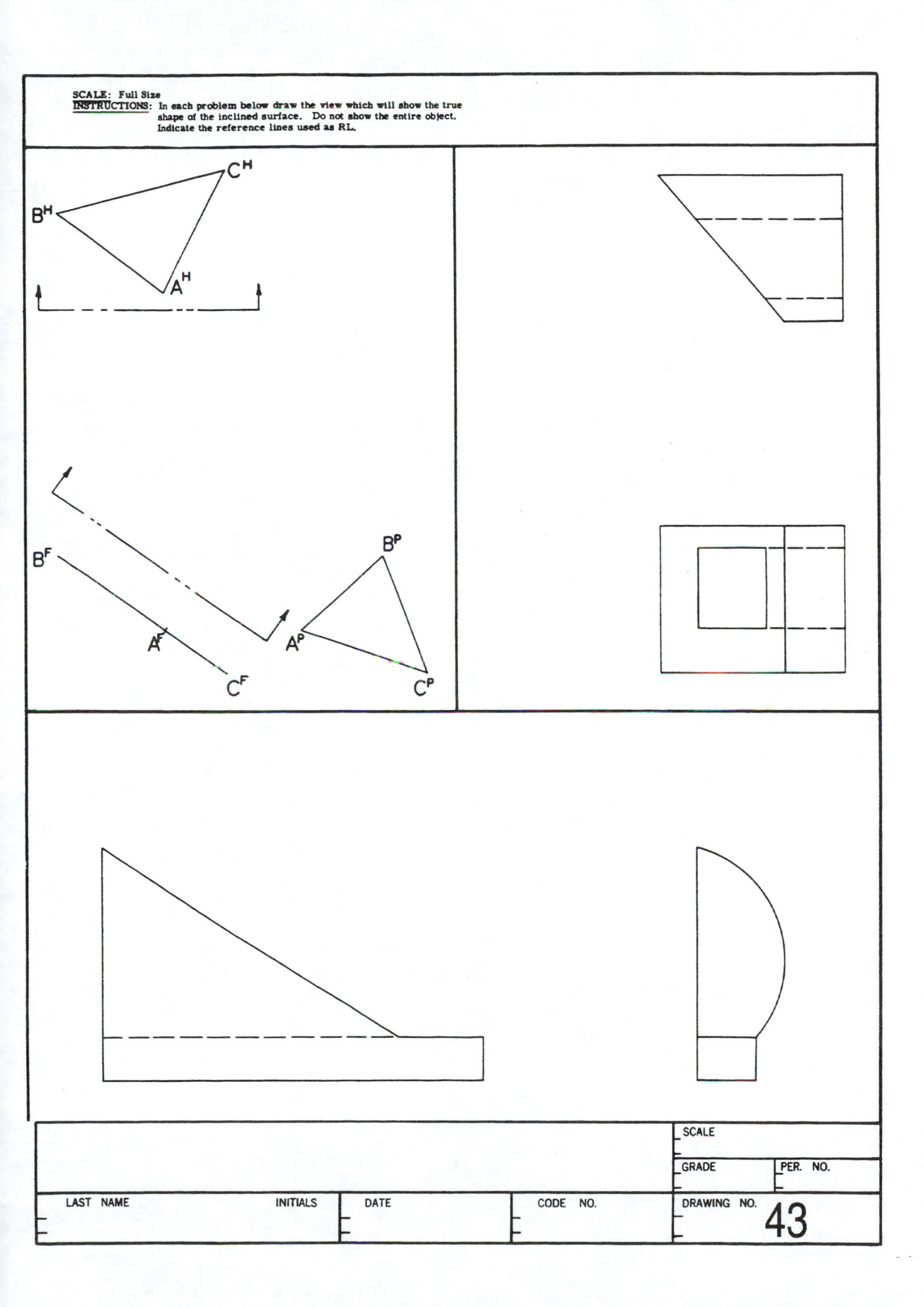

INSTRUCTIONS: Show all reference planes and directional arrows to indicate equivalent distances as required by the instructor. Note: All solutions must be obtained by projection. Label all edge views as E.V., and all true length lines as T.L.

Prob. 1. Draw the partial view which shows the true shape of the inclined surface.

Prob. 2. Add notations to top and side views, then draw the partial view which shows the true shape of surface ABC.

Prob. 3. Draw the partial view which shows the true shape of the inclined surface.

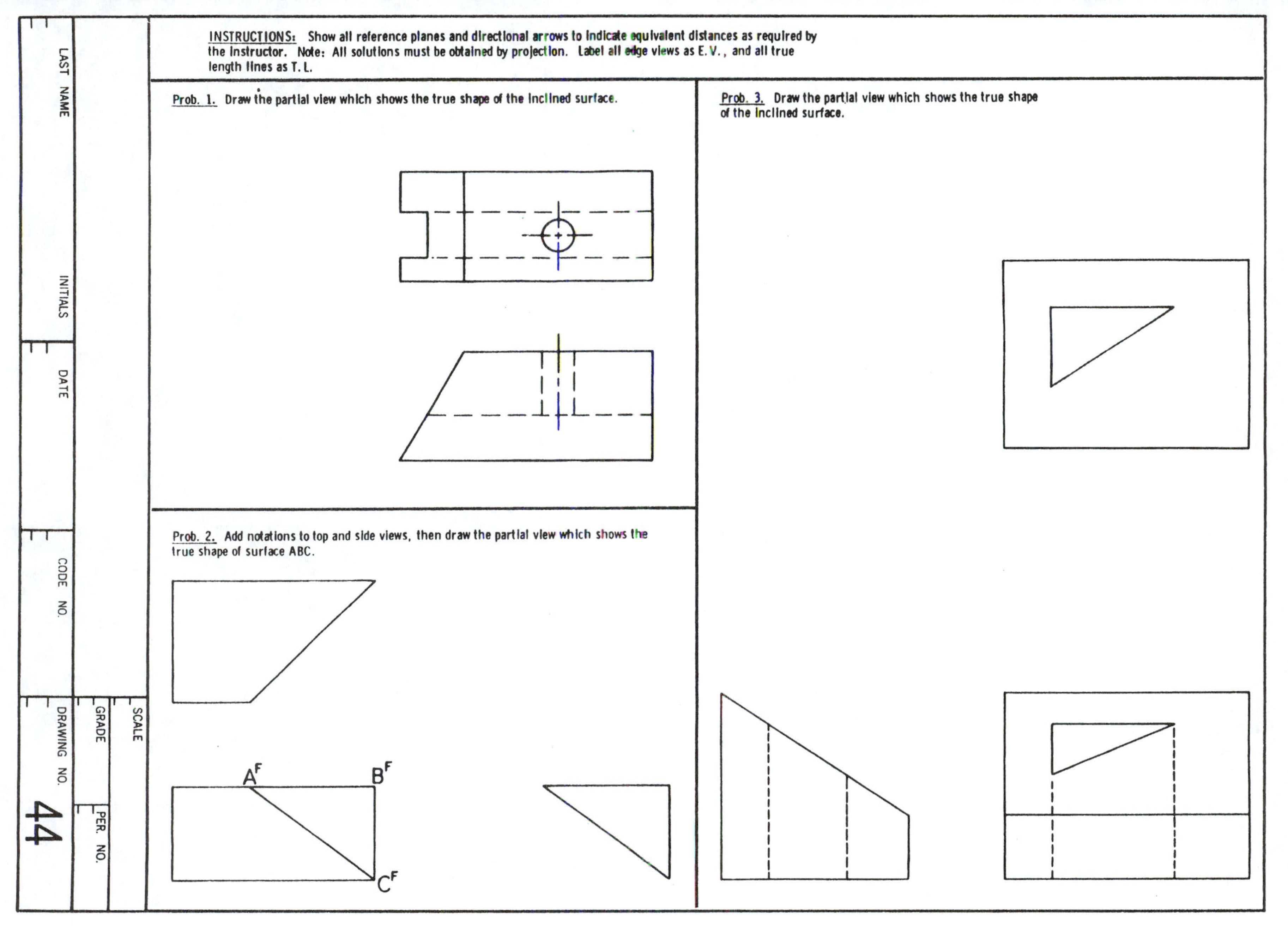

DESCRIPTIVE GEOMETRY

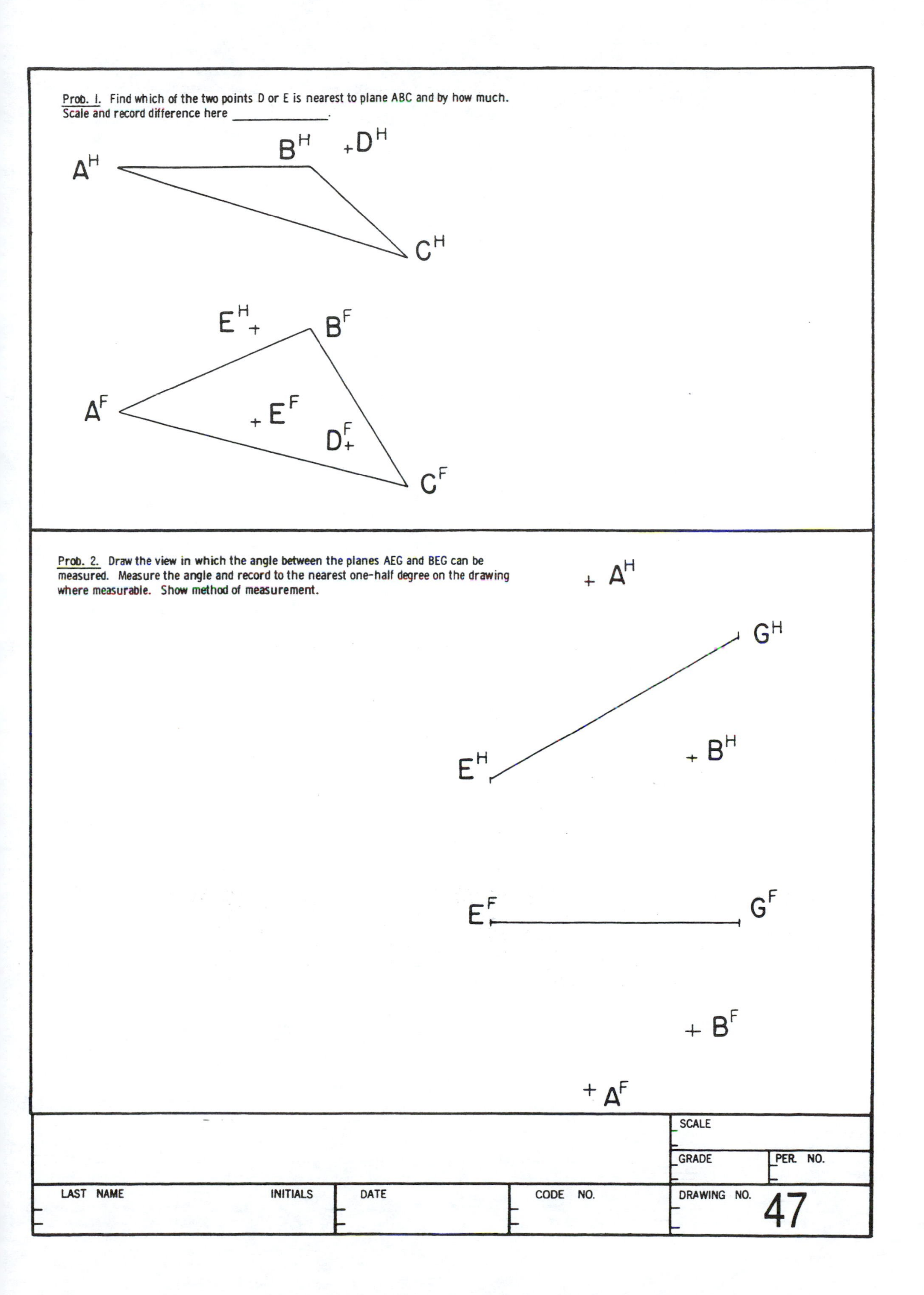
Prob. I. Find which of the two points D or E is nearest to plane ABC and by how much.
Scale and record difference here ______________.
A^H
B^H
D^H
C^H
E^H
B^F
A^F
E^F
D^F
C^F
Prob. 2. Draw the view in which the angle between the planes AEG and BEG can be measured. Measure the angle and record to the nearest one-half degree on the drawing where measurable. Show method of measurement.
A^H
G^H
E^H
B^H
E^F
G^F
B^F
A^F
SCALE
GRADE
PER. NO.
LAST NAME
INITIALS
DATE
CODE NO.
DRAWING NO.
47

INSTRUCTIONS: The two given views represent a center line of a tube EF welded to a plate ABCD in an experimental mock-up. A production drawing of this element must show, among other things, the true angle between the tube and plate. Using appropriately projected auxiliary views only, draw the view which shows the required angle. Determine and record the angle, to the nearest degree where it is measurable.

B^H C^H F^H E^H D^H A^H

A^F B^F E^F F^F D^F C^F

LAST NAME	INITIALS	DATE	CODE NO.	SCALE / GRADE / PER. NO. / DRAWING NO. 48

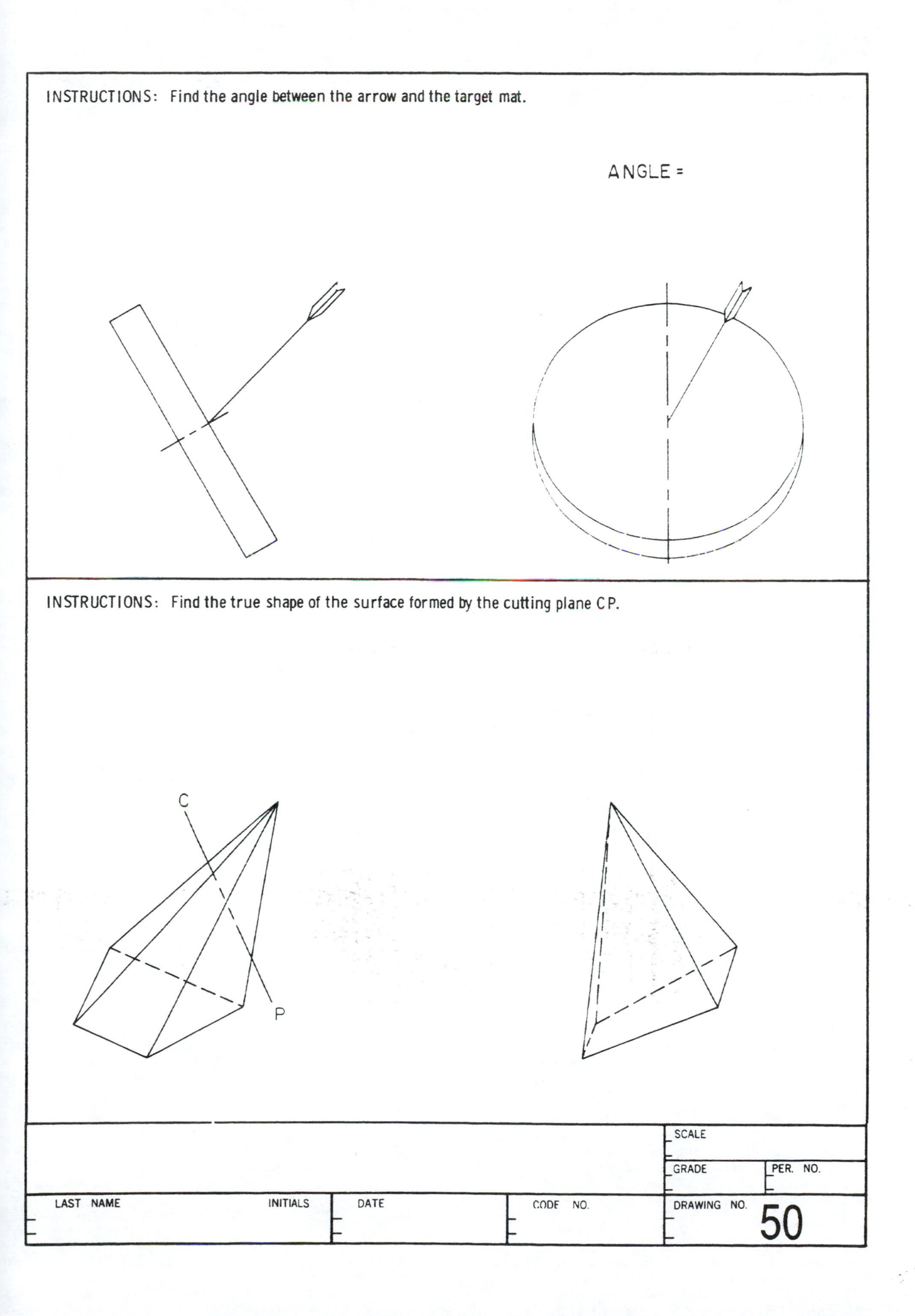
INSTRUCTIONS: Find the angle between the arrow and the target mat.
ANGLE =
INSTRUCTIONS: Find the true shape of the surface formed by the cutting plane CP.
C
P
SCALE
GRADE
PER. NO.
LAST NAME
INITIALS
DATE
CODE NO.
DRAWING NO.
50

INSTRUCTIONS: Two views of a section of sheet metal duct are given. Before the sheet metal flat pattern can be cut to the proper length, it is necessary to have a view which shows the true perimeter of the duct. This is found in the view which shows the true cross-sectional shape. Using appropriately projected auxiliary views only, find the required view.

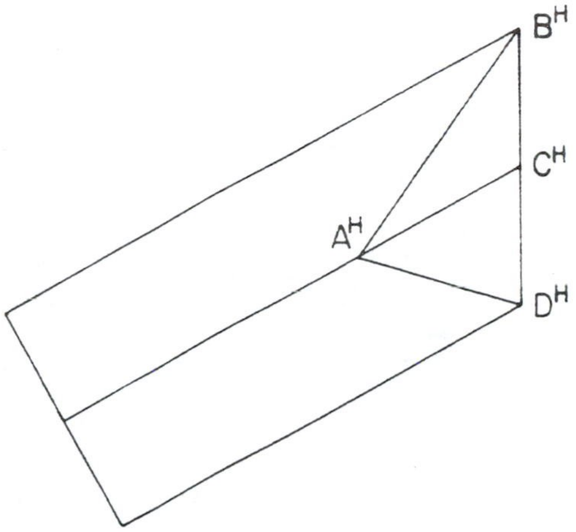

B^H

C^H

A^H

D^H

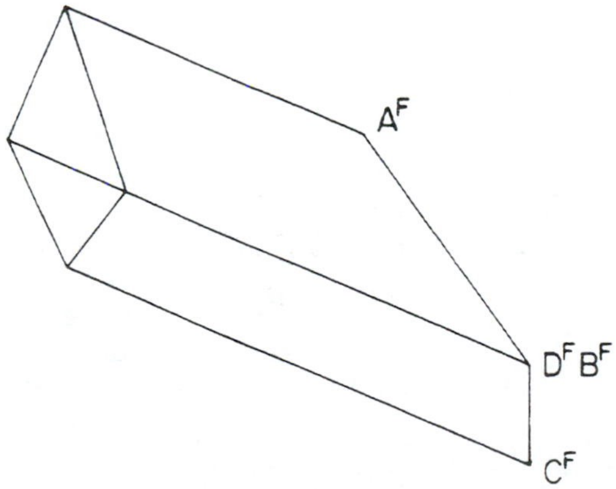

A^F

$D^F B^F$

C^F

LAST NAME	INITIALS	DATE	CODE NO.	SCALE	
				GRADE	PER. NO.
				DRAWING NO.	51

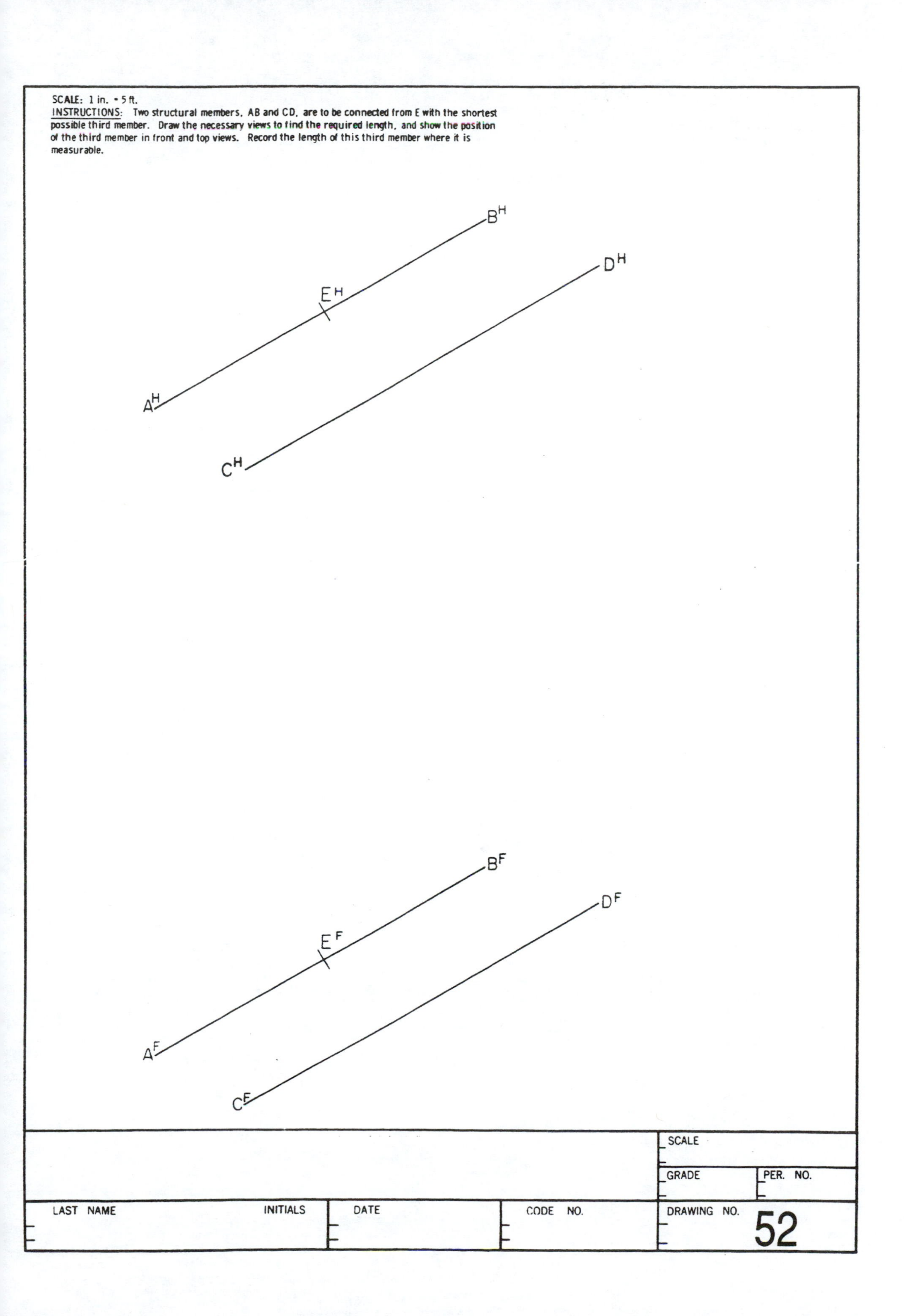
SCALE: 1 in. = 5 ft.
INSTRUCTIONS: Two structural members, AB and CD, are to be connected from E with the shortest possible third member. Draw the necessary views to find the required length, and show the position of the third member in front and top views. Record the length of this third member where it is measurable.
B H
D H
E H
A H
C H
B F
D F
E F
A F
C F
SCALE
GRADE
PER. NO.
LAST NAME
INITIALS
DATE
CODE NO.
DRAWING NO.
52

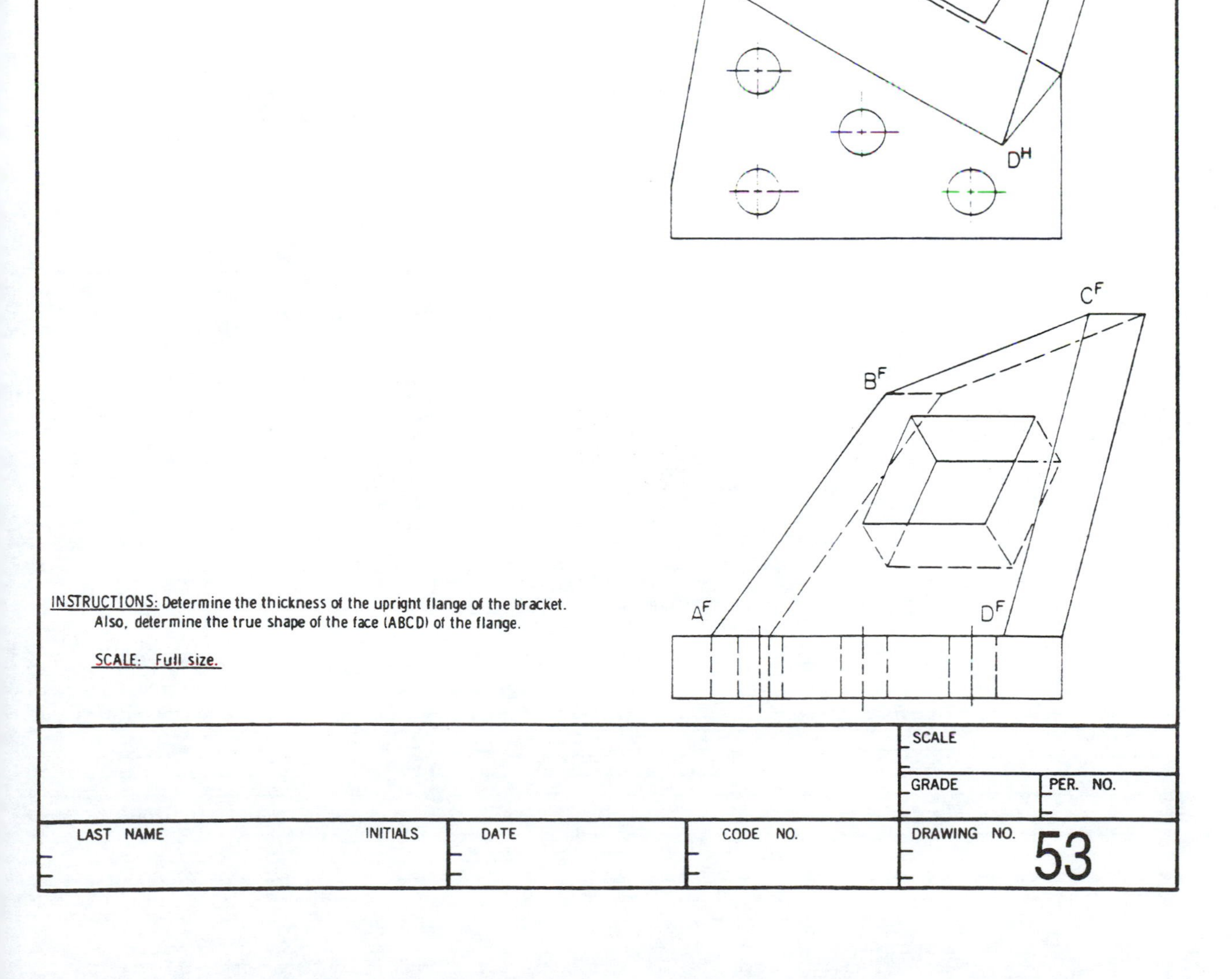

INSTRUCTIONS: Determine the thickness of the upright flange of the bracket.
Also, determine the true shape of the face (ABCD) of the flange.

SCALE: Full size.

			SCALE	
			GRADE	PER. NO.
LAST NAME INITIALS	DATE	CODE NO.	DRAWING NO.	53

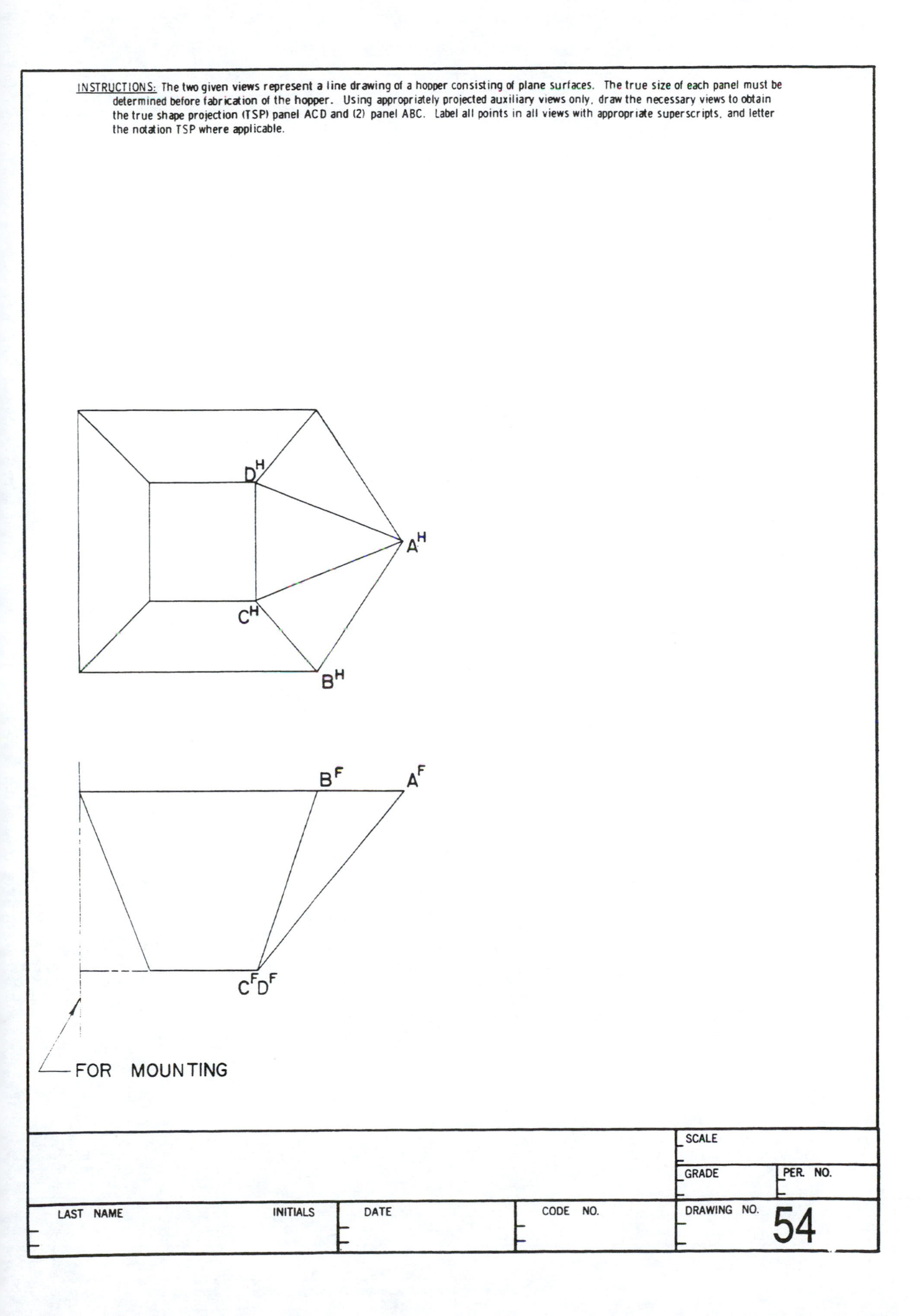

INSTRUCTIONS: The two given views represent a line drawing of a hopper consisting of plane surfaces. The true size of each panel must be determined before fabrication of the hopper. Using appropriately projected auxiliary views only, draw the necessary views to obtain the true shape projection (TSP) panel ACD and (2) panel ABC. Label all points in all views with appropriate superscripts, and letter the notation TSP where applicable.
D^H
A^H
C^H
B^H
B^F
A^F
C^F D^F
FOR MOUNTING
SCALE
GRADE
PER. NO.
LAST NAME
INITIALS
DATE
CODE NO.
DRAWING NO.
54

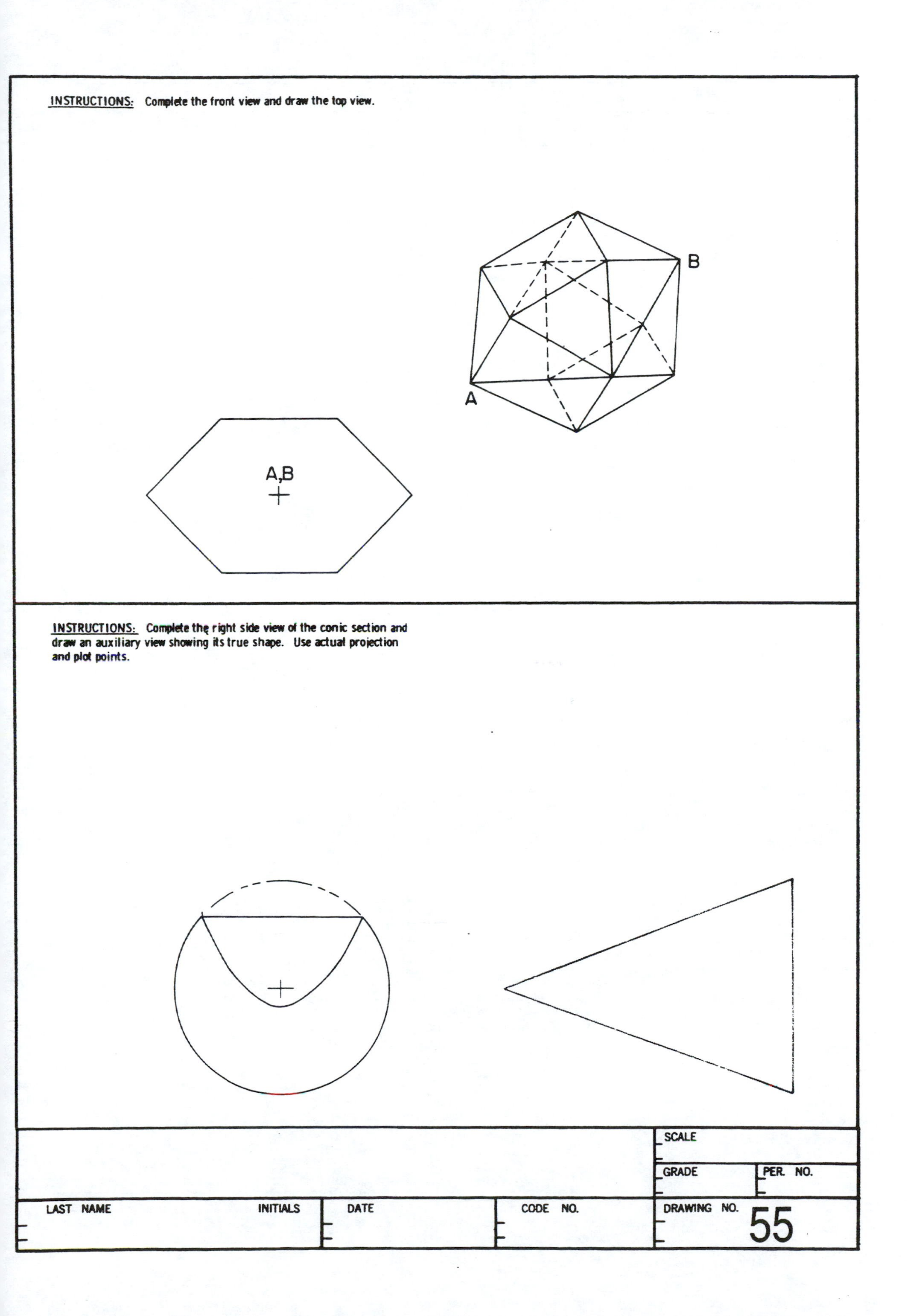
INSTRUCTIONS: Complete the front view and draw the top view.
B
A
A,B
INSTRUCTIONS: Complete the right side view of the conic section and draw an auxiliary view showing its true shape. Use actual projection and plot points.
SCALE
GRADE
PER. NO.
LAST NAME
INITIALS
DATE
CODE NO.
DRAWING NO.
55

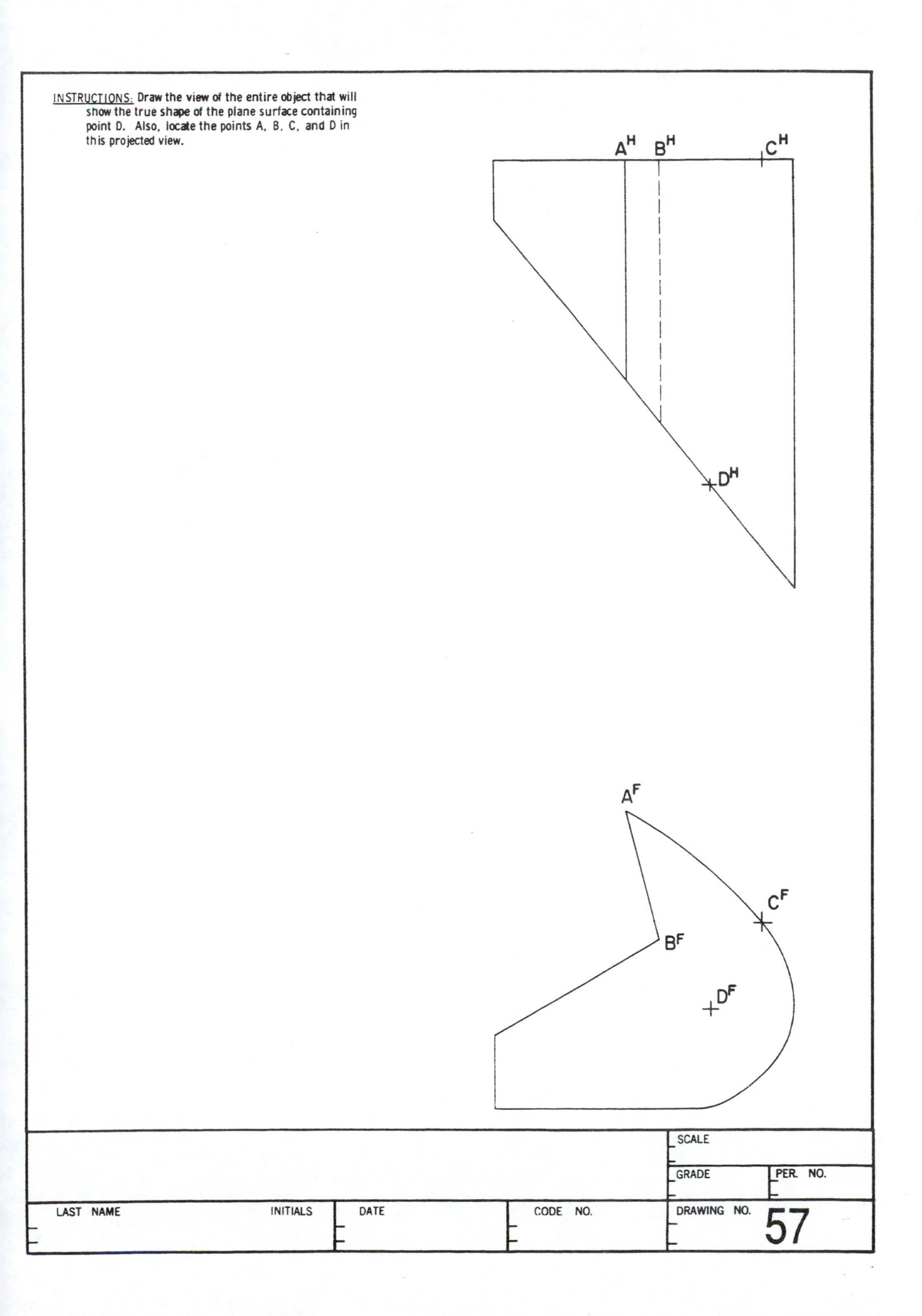
INSTRUCTIONS: Draw the view of the entire object that will show the true shape of the plane surface containing point D. Also, locate the points A, B, C, and D in this projected view.
A^H
B^H
C^H
D^H
A^F
C^F
B^F
D^F
SCALE
GRADE
PER. NO.
LAST NAME
INITIALS
DATE
CODE NO.
DRAWING NO.
57

SECTION VIEWS

INSTRUCTIONS: The given pictorial shows a full section. Using the completed top view, the given dimensions, and the pictorial make a full sectional view of the object. Prior to drawing the sectional view, locate the cutting plane in the top view and add necessary center lines.

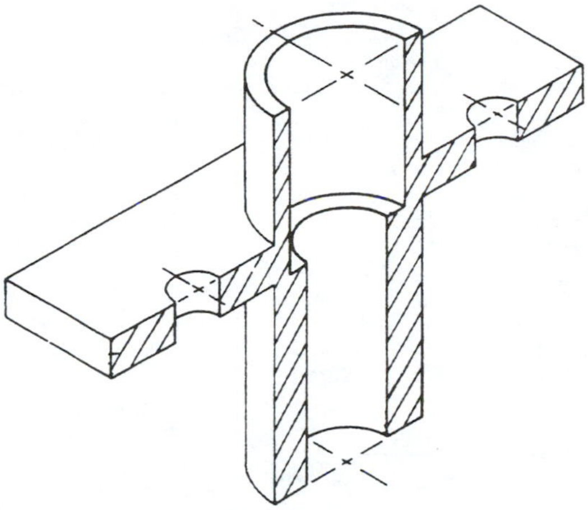

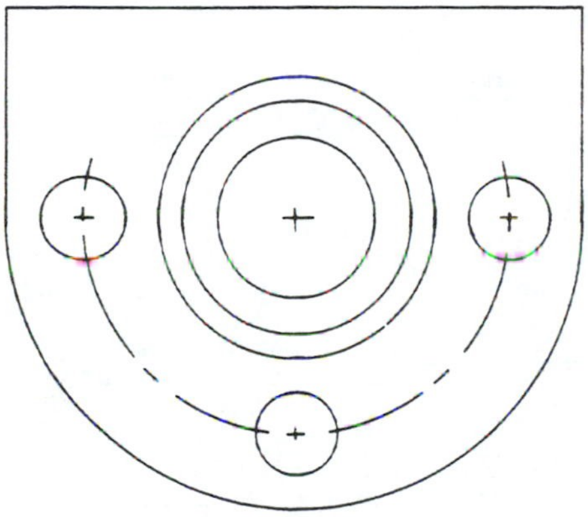

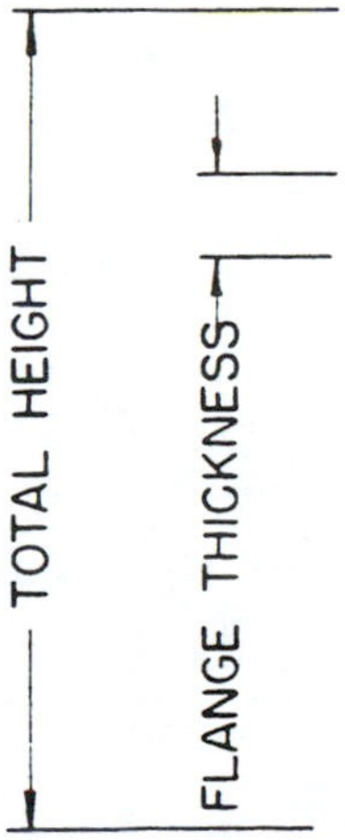

TOTAL HEIGHT

FLANGE THICKNESS

LAST NAME	INITIALS	DATE	CODE NO.	SCALE
				GRADE / PER. NO.
				DRAWING NO. 58

INSTRUCTIONS: If true projection were to be used, the usual views of some objects would be less easily interpreted because of the placement of some portions or details of the object. Hence, these details, such as bosses, lugs, ribs, and holes are revolved for the sake of clarity. This procedure is known as the "Treatment of Radial Features."

1) Given is the complete top view of a Cap and a partial front view to show the height of several features such as lugs, bosses, etc.

2) Using the information furnished by the given views, make a full sectional view. Follow the recommendations given for treatment of radial features.

3) After the sectional view is completed, identify the cut of the section with a cutting plane line and letters.

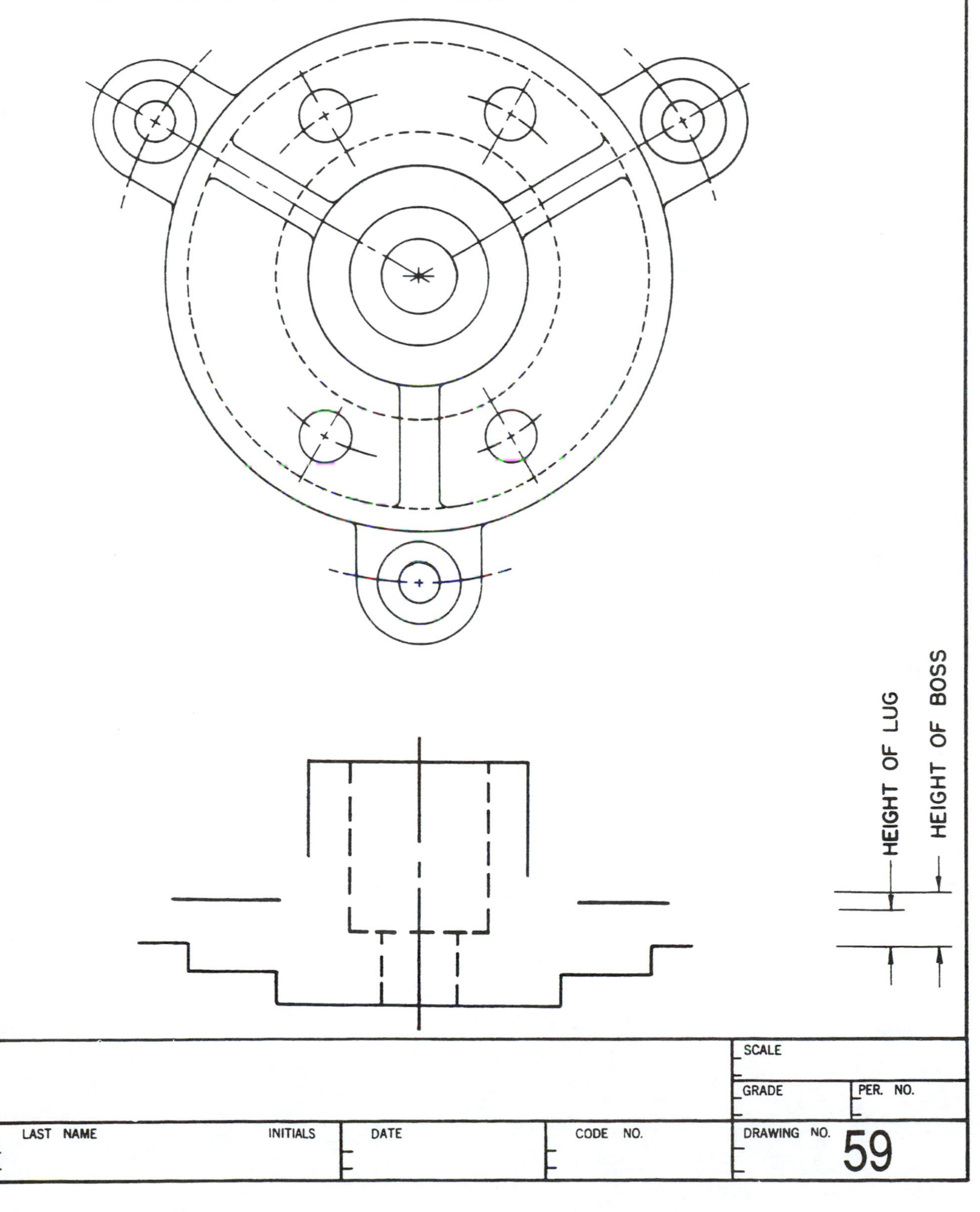

	SCALE	
	GRADE	PER. NO.
LAST NAME INITIALS | DATE | CODE NO.	DRAWING NO.	59

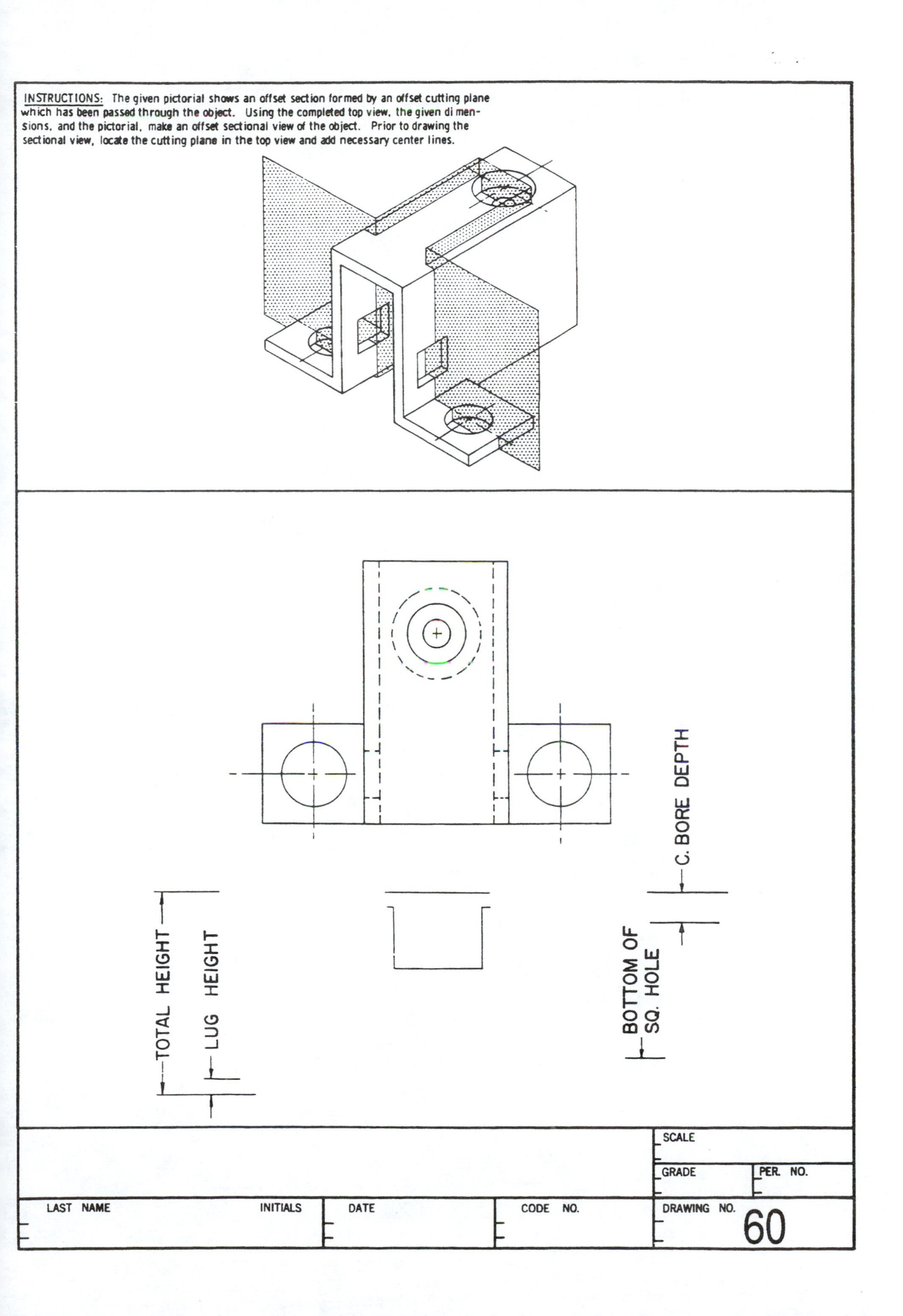
INSTRUCTIONS: The given pictorial shows an offset section formed by an offset cutting plane which has been passed through the object. Using the completed top view, the given dimensions, and the pictorial, make an offset sectional view of the object. Prior to drawing the sectional view, locate the cutting plane in the top view and add necessary center lines.
TOTAL HEIGHT
LUG HEIGHT
C. BORE DEPTH
BOTTOM OF SQ. HOLE
SCALE
GRADE
PER. NO.
LAST NAME
INITIALS
DATE
CODE NO.
DRAWING NO.
60

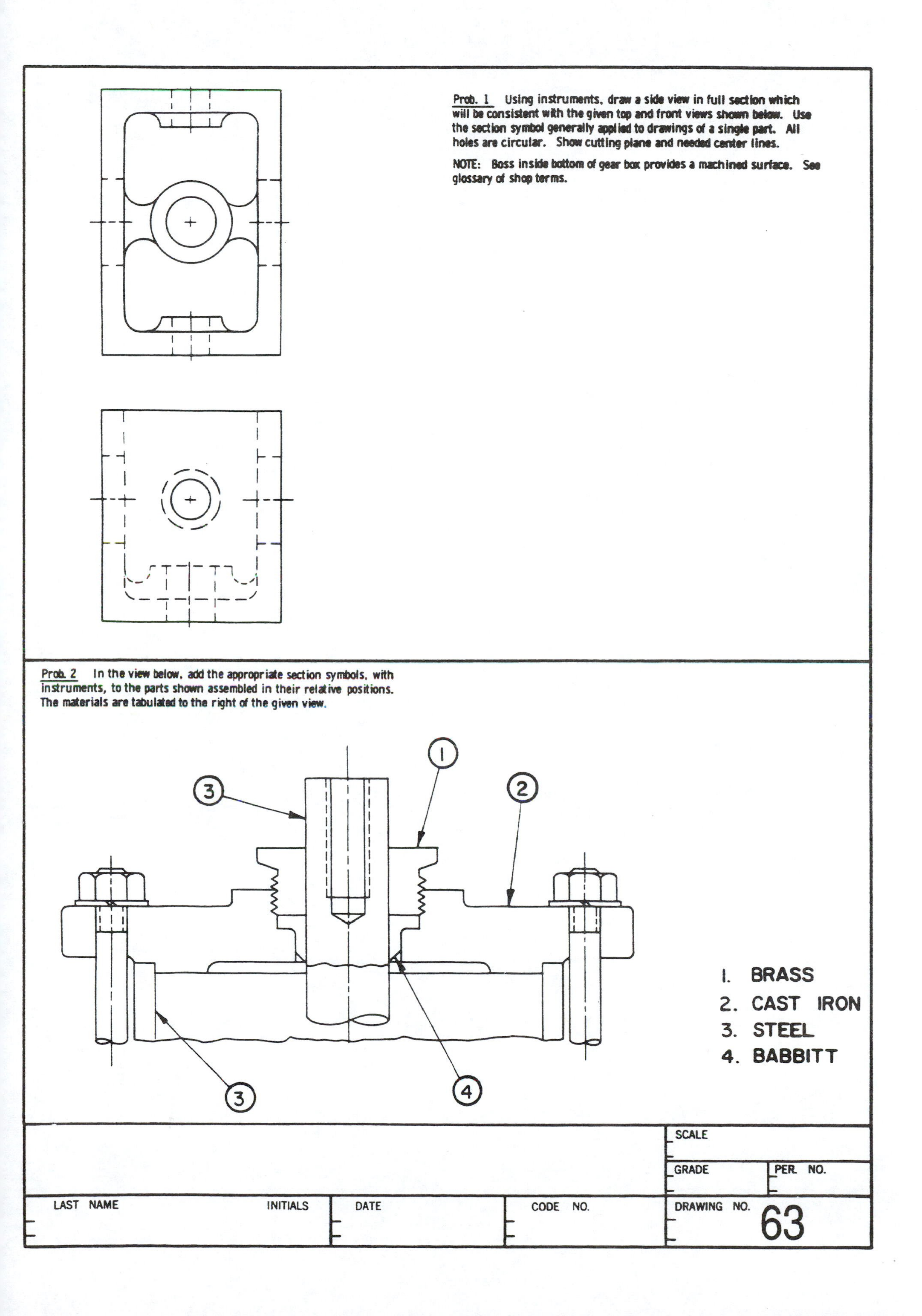
Prob. 1 Using instruments, draw a side view in full section which will be consistent with the given top and front views shown below. Use the section symbol generally applied to drawings of a single part. All holes are circular. Show cutting plane and needed center lines.
NOTE: Boss inside bottom of gear box provides a machined surface. See glossary of shop terms.
Prob. 2 In the view below, add the appropriate section symbols, with instruments, to the parts shown assembled in their relative positions. The materials are tabulated to the right of the given view.
1
2
3
3
4
1. BRASS
2. CAST IRON
3. STEEL
4. BABBITT
SCALE
GRADE
PER. NO.
LAST NAME
INITIALS
DATE
CODE NO.
DRAWING NO.
63

DIMENSIONING

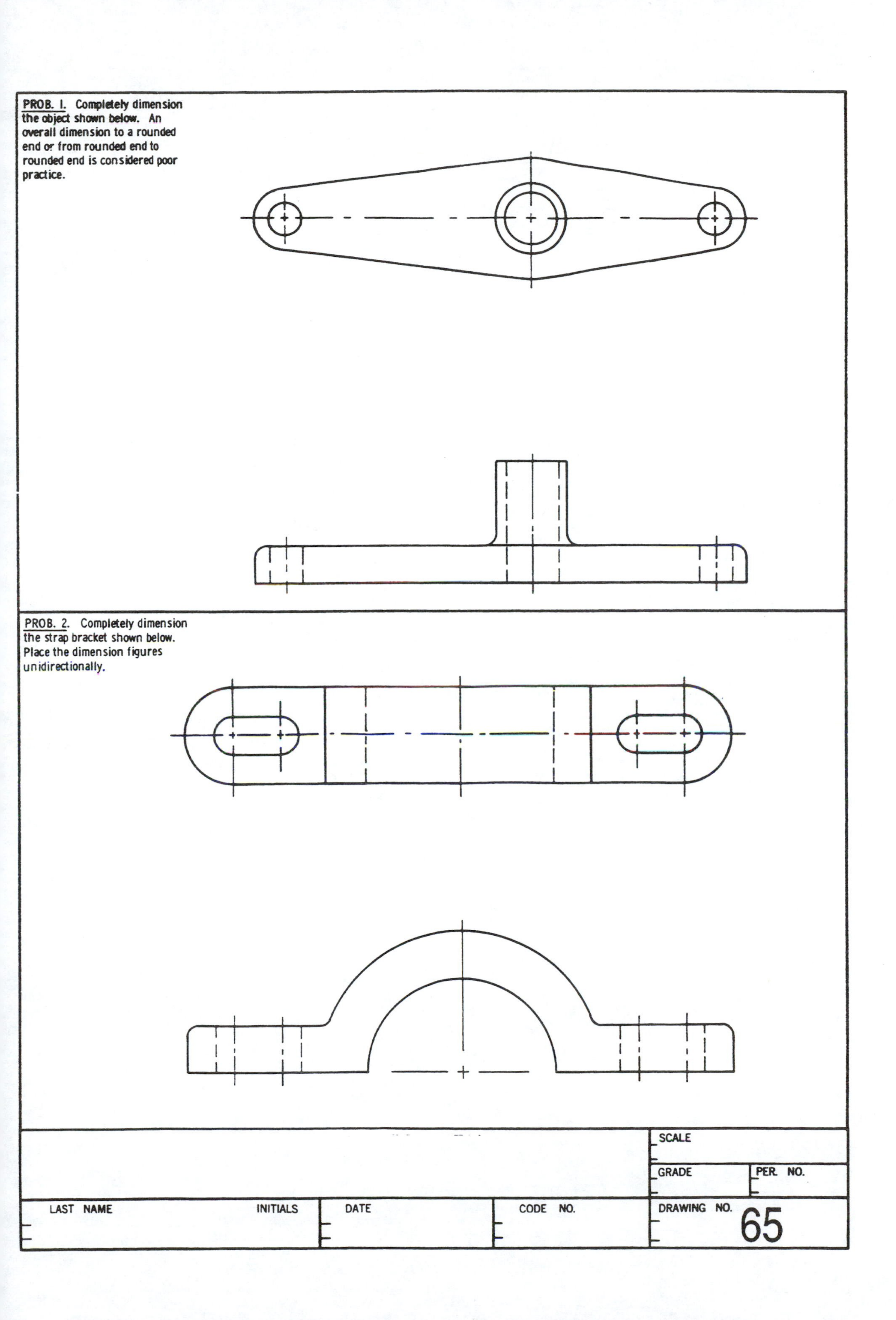
PROB. 1. Completely dimension the object shown below. An overall dimension to a rounded end or from rounded end to rounded end is considered poor practice.
PROB. 2. Completely dimension the strap bracket shown below. Place the dimension figures unidirectionally.
SCALE
GRADE
PER. NO.
LAST NAME
INITIALS
DATE
CODE NO.
DRAWING NO.
65

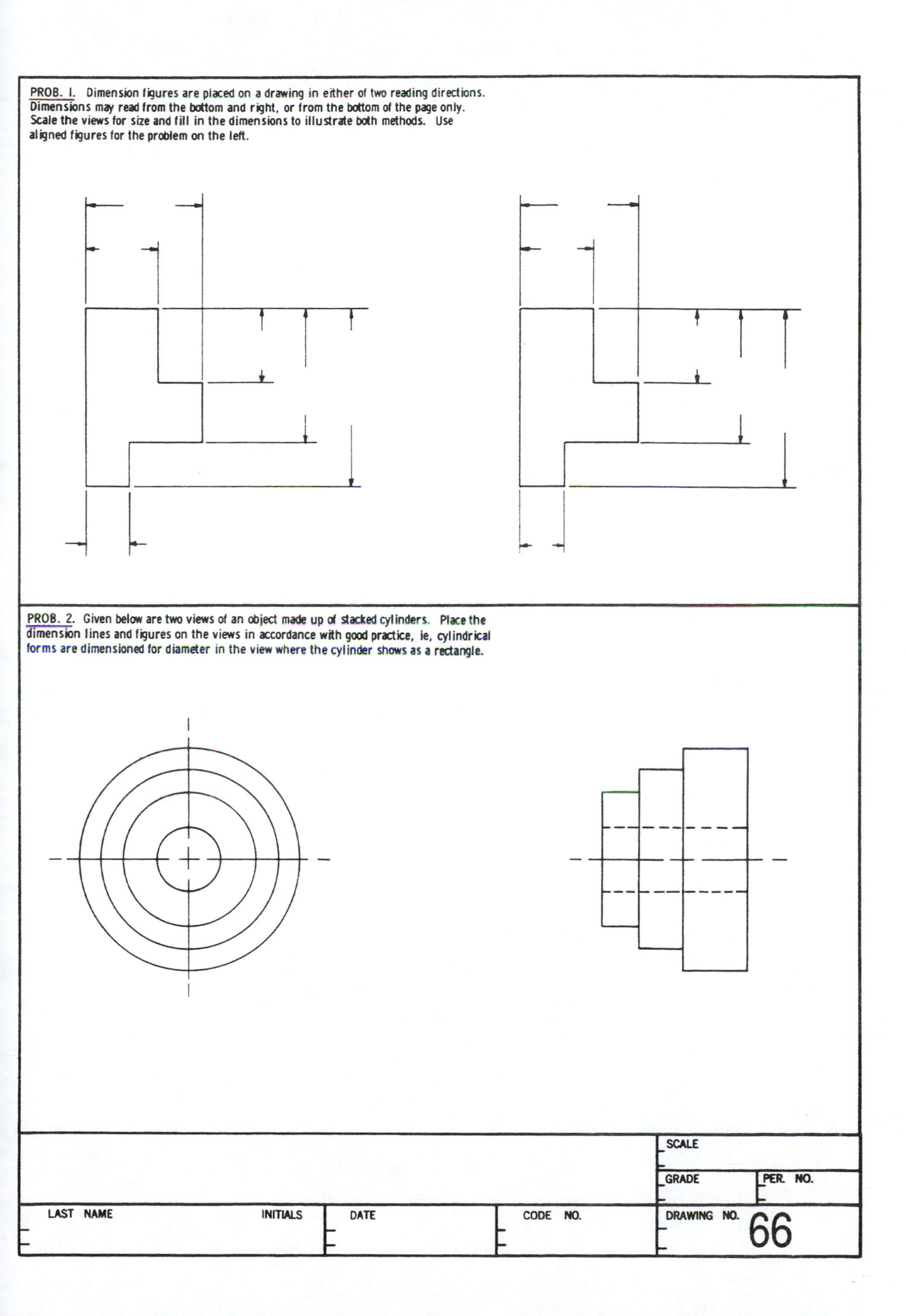
PROB. 1. Dimension figures are placed on a drawing in either of two reading directions. Dimensions may read from the bottom and right, or from the bottom of the page only. Scale the views for size and fill in the dimensions to illustrate both methods. Use aligned figures for the problem on the left.
PROB. 2. Given below are two views of an object made up of stacked cylinders. Place the dimension lines and figures on the views in accordance with good practice, ie, cylindrical forms are dimensioned for diameter in the view where the cylinder shows as a rectangle.
SCALE
GRADE
PER. NO.
LAST NAME
INITIALS
DATE
CODE NO.
DRAWING NO.
66

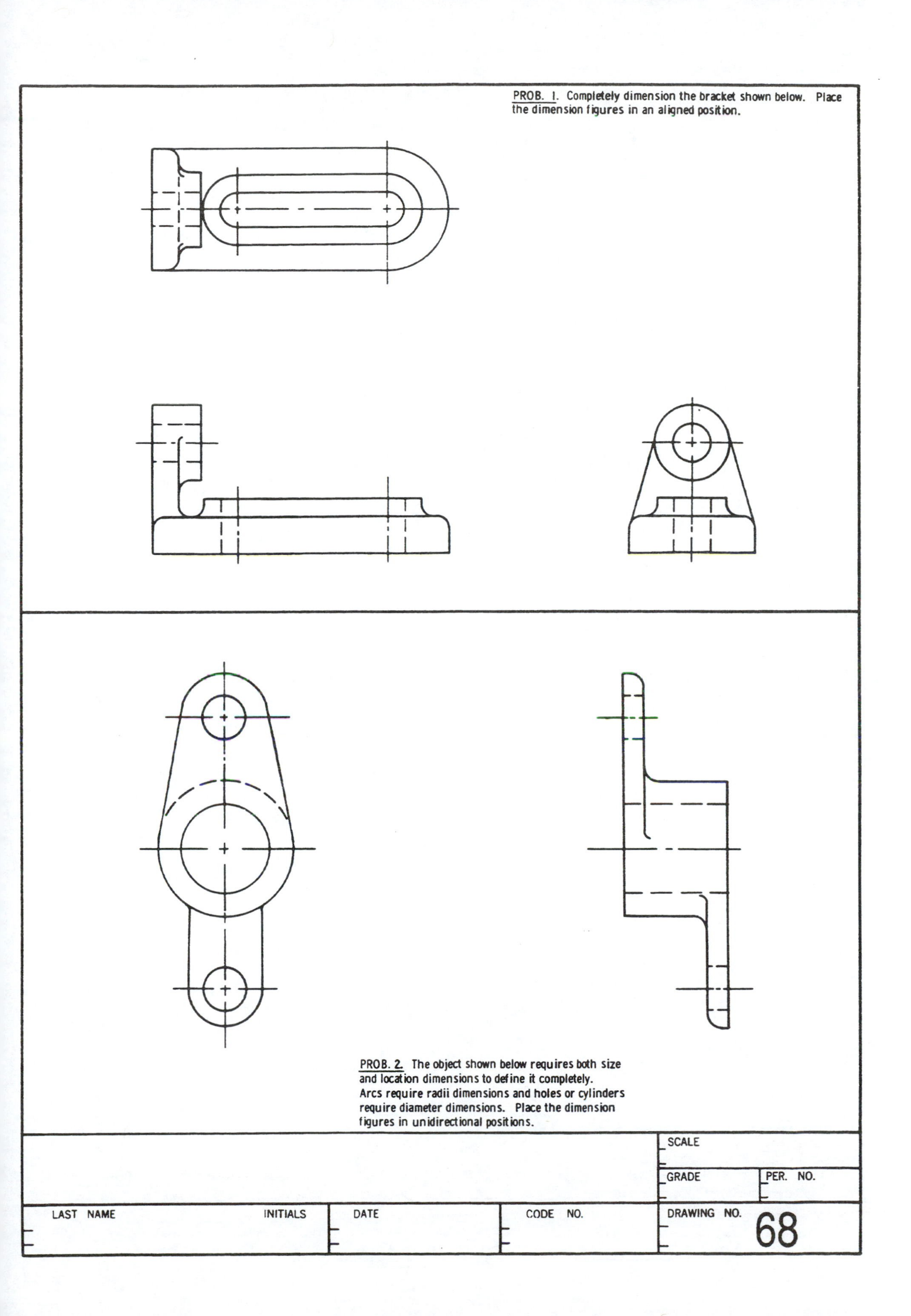
PROB. 1. Completely dimension the bracket shown below. Place the dimension figures in an aligned position.
PROB. 2. The object shown below requires both size and location dimensions to define it completely. Arcs require radii dimensions and holes or cylinders require diameter dimensions. Place the dimension figures in unidirectional positions.
SCALE
GRADE
PER. NO.
LAST NAME
INITIALS
DATE
CODE NO.
DRAWING NO.
68

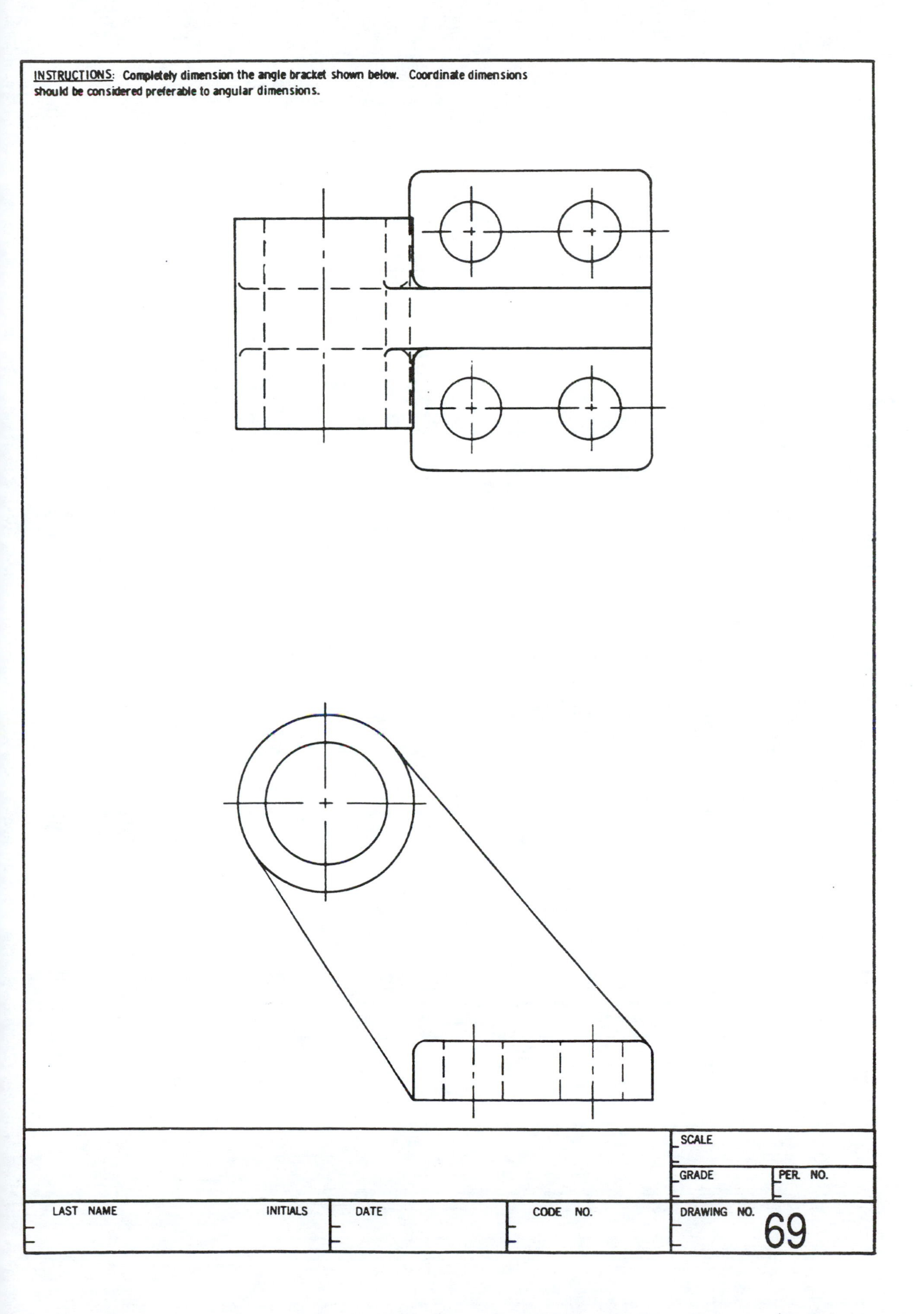
INSTRUCTIONS: Completely dimension the angle bracket shown below. Coordinate dimensions should be considered preferable to angular dimensions.
SCALE
GRADE
PER. NO.
LAST NAME
INITIALS
DATE
CODE NO.
DRAWING NO.
69

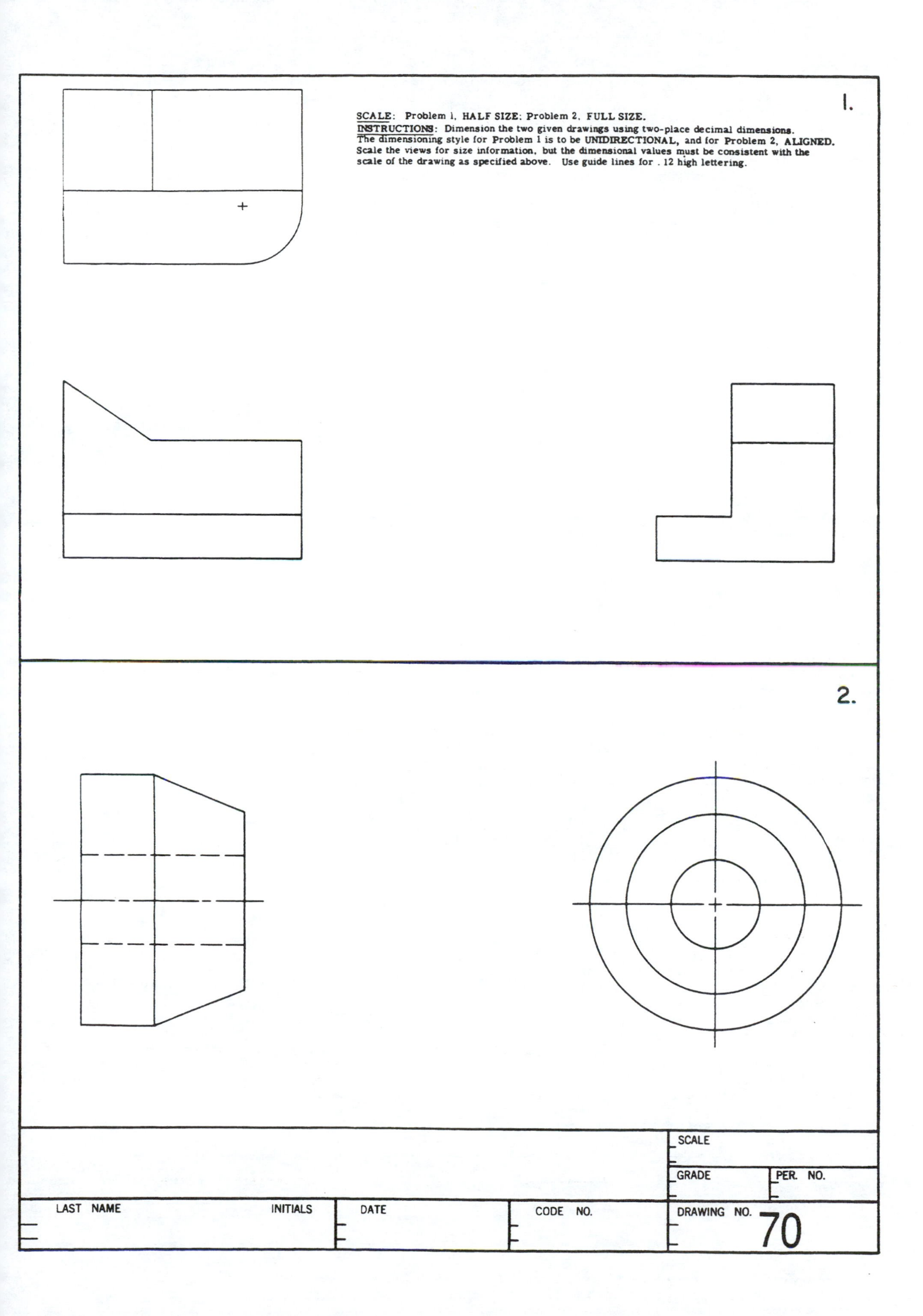
1.
SCALE: Problem 1, HALF SIZE; Problem 2, FULL SIZE.
INSTRUCTIONS: Dimension the two given drawings using two-place decimal dimensions.
The dimensioning style for Problem 1 is to be UNIDIRECTIONAL, and for Problem 2, ALIGNED.
Scale the views for size information, but the dimensional values must be consistent with the
scale of the drawing as specified above. Use guide lines for .12 high lettering.
2.
SCALE
GRADE
PER. NO.
LAST NAME
INITIALS
DATE
CODE NO.
DRAWING NO.
70

Completely dimension the clevis shown below. Use decimals in unidirectional positions.

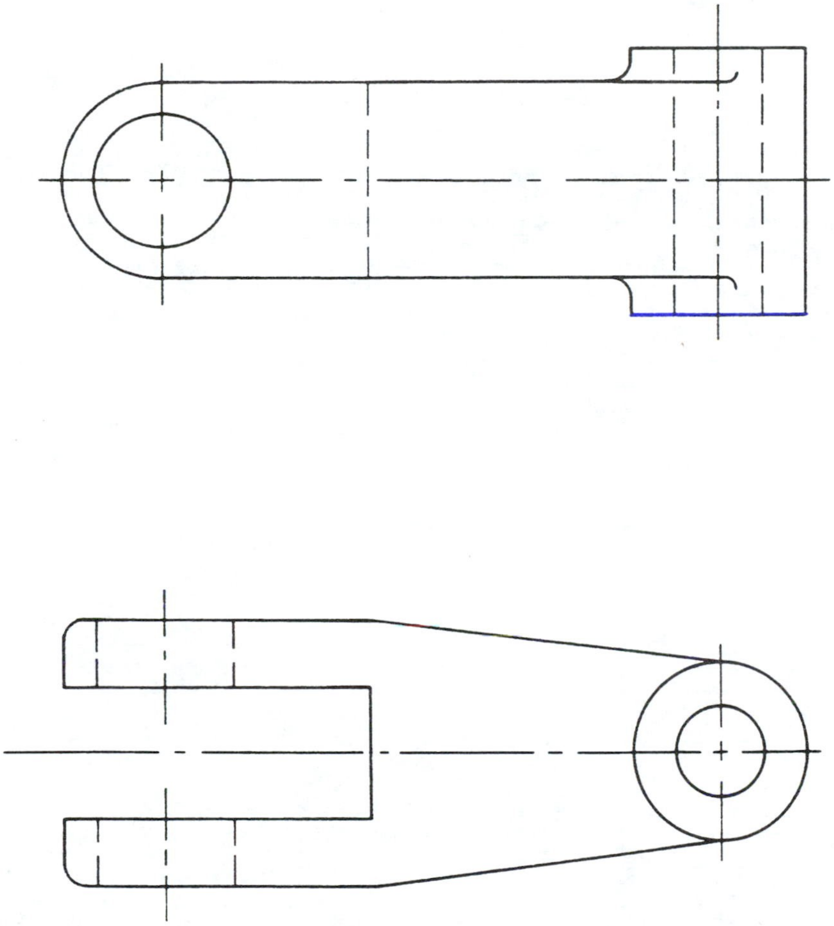

	SCALE	
	GRADE	PER. NO.

LAST NAME INITIALS	DATE	CODE NO.	DRAWING NO. 72

INSTRUCTIONS: The location dimensions for establishing the positioning of holes may be given in accordance with the so-called coordinate method. Angular location dimensions and dimensions of the non-coordinate form may be used in addition as reference dimensions. Dimension completely the drawing below using size and location dimensions and reference dimensions where necessary.

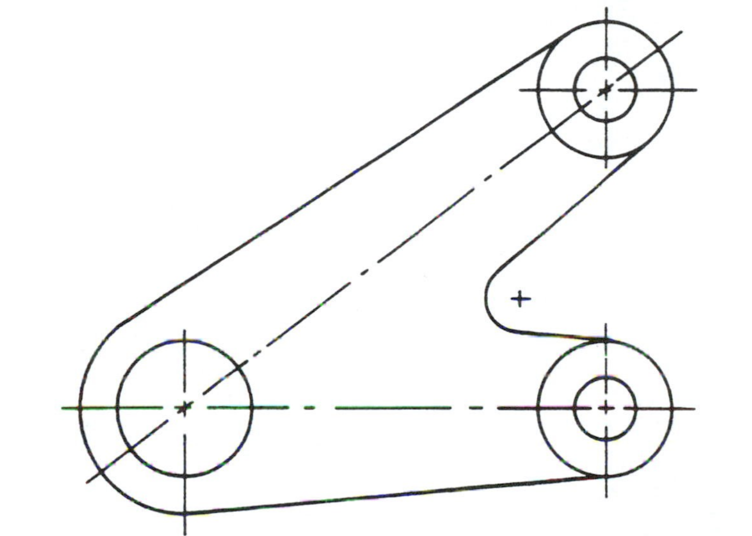

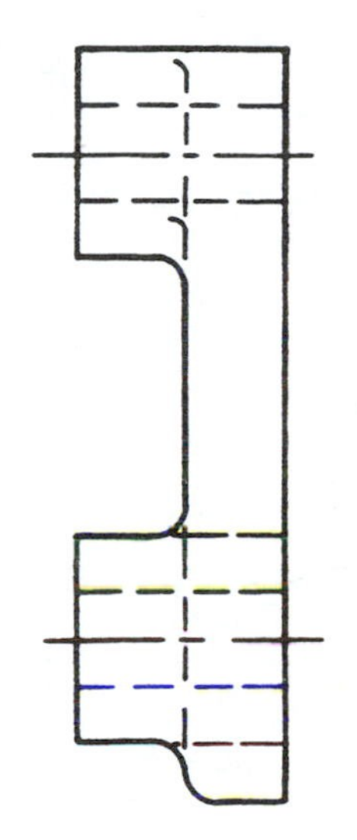

	SCALE	
	GRADE	PER. NO.

LAST NAME INITIALS	DATE	CODE NO.	DRAWING NO. 73

INSTRUCTIONS: Completely dimension the object shown. Be sure that all component shapes are dimensioned for both location and size. All through holes are the same size and all rounded corners have the same radii. Scale the views for size information and use two-place decimals placed unidirectionally.

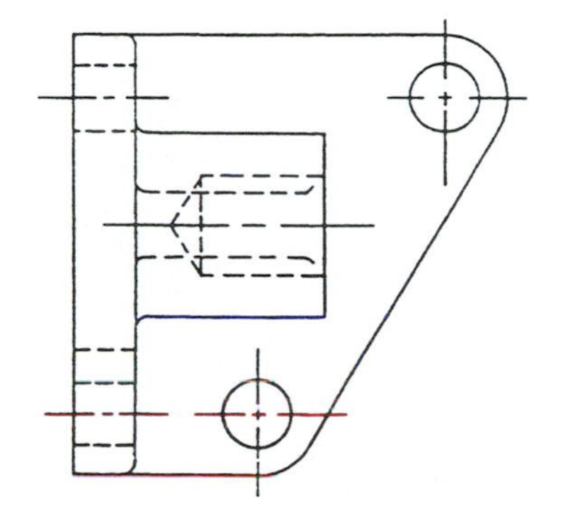

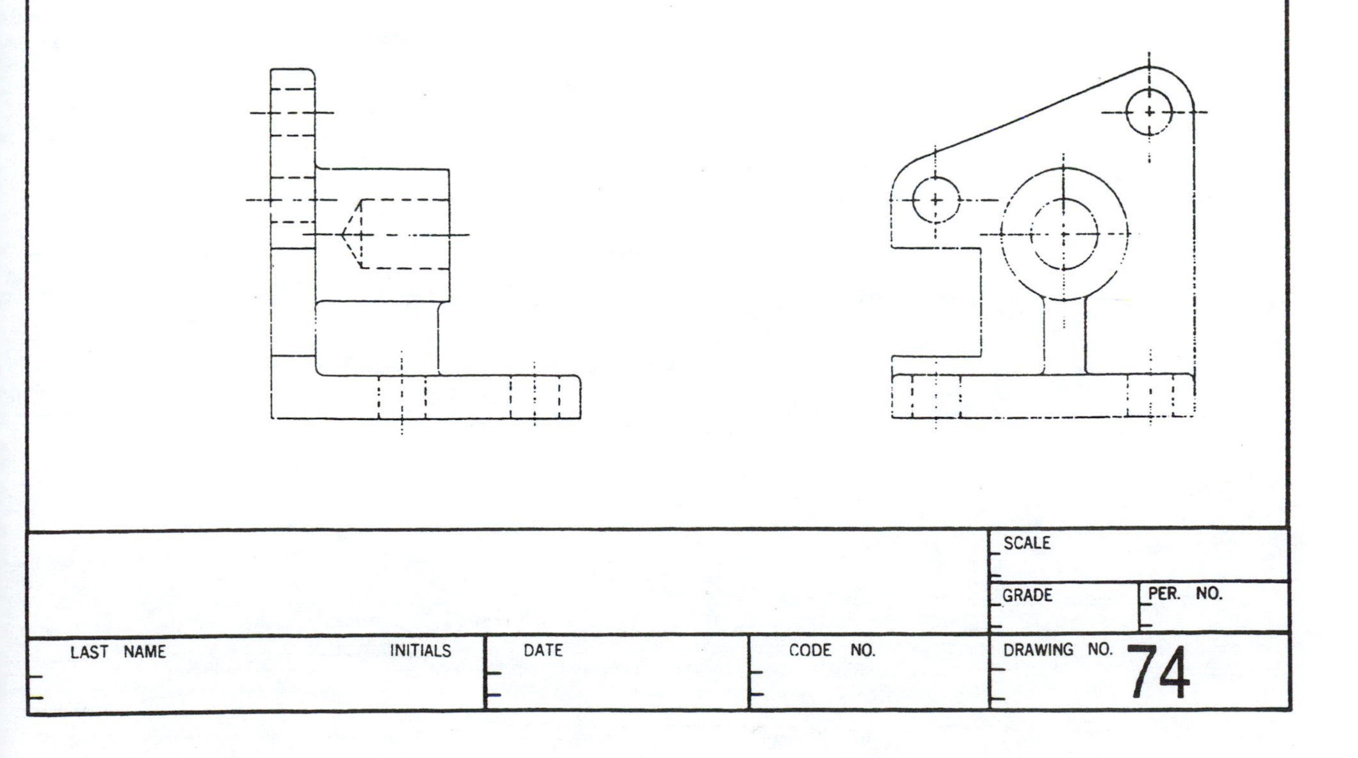

			SCALE	
			GRADE	PER. NO.
LAST NAME INITIALS	DATE	CODE NO.	DRAWING NO. 74	

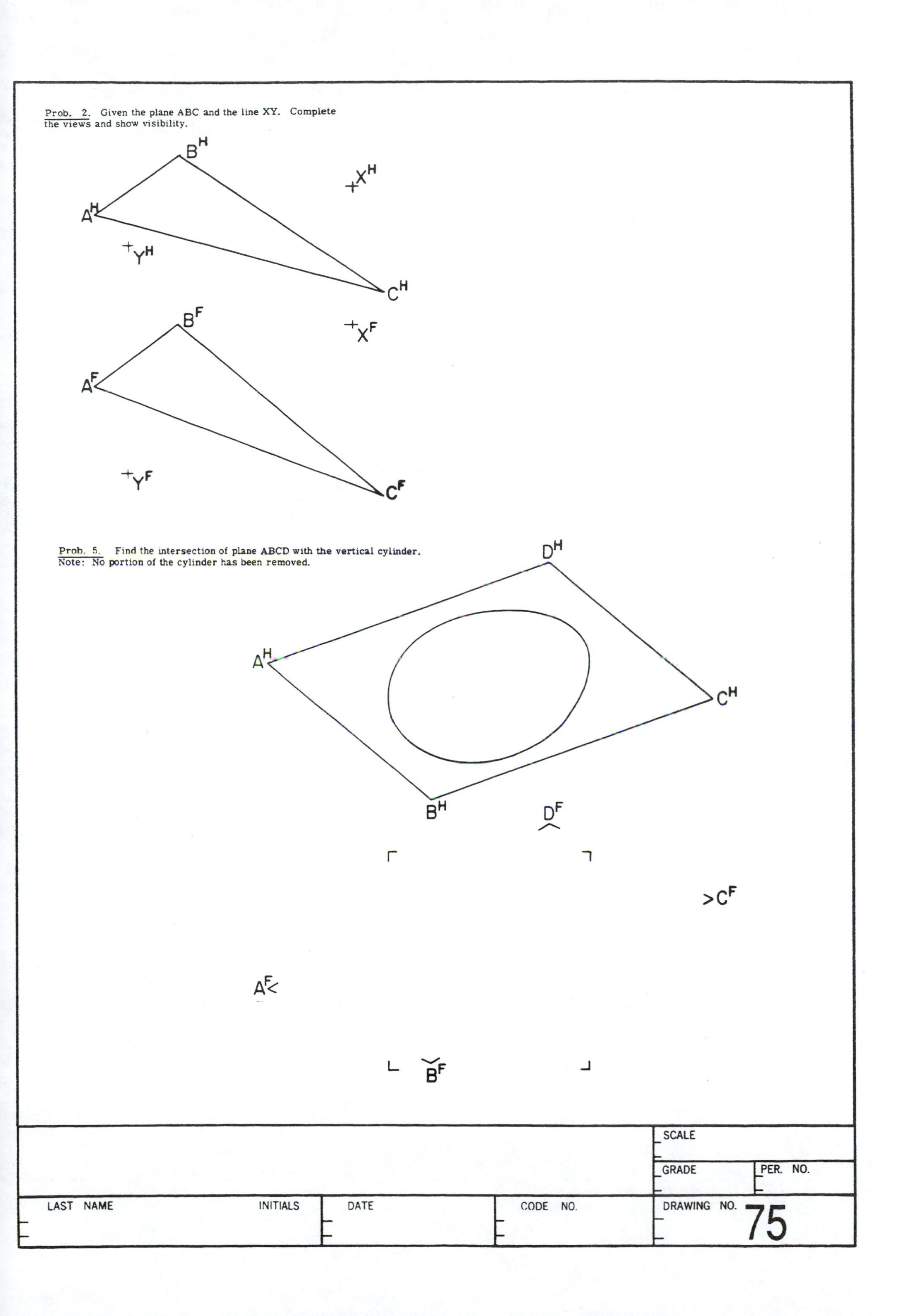
Prob. 2. Given the plane ABC and the line XY. Complete the views and show visibility.
B^H
X^H
A^H
Y^H
C^H
B^F
X^F
A^F
Y^F
C^F
Prob. 5. Find the intersection of plane ABCD with the vertical cylinder.
Note: No portion of the cylinder has been removed.
D^H
A^H
C^H
B^H
D^F
C^F
A^F
B^F
SCALE
GRADE
PER. NO.
LAST NAME
INITIALS
DATE
CODE NO.
DRAWING NO.
75

FASTENERS

DIMENSIONING PRACTICES - SHOP NOTES

INSTRUCTIONS: Given below are several drawings of parts with holes for which the draftsman must specify size and shop operations. Letter the correct note for each hole using the guide lines provided. Needed size dimensions can be determined by scaling.

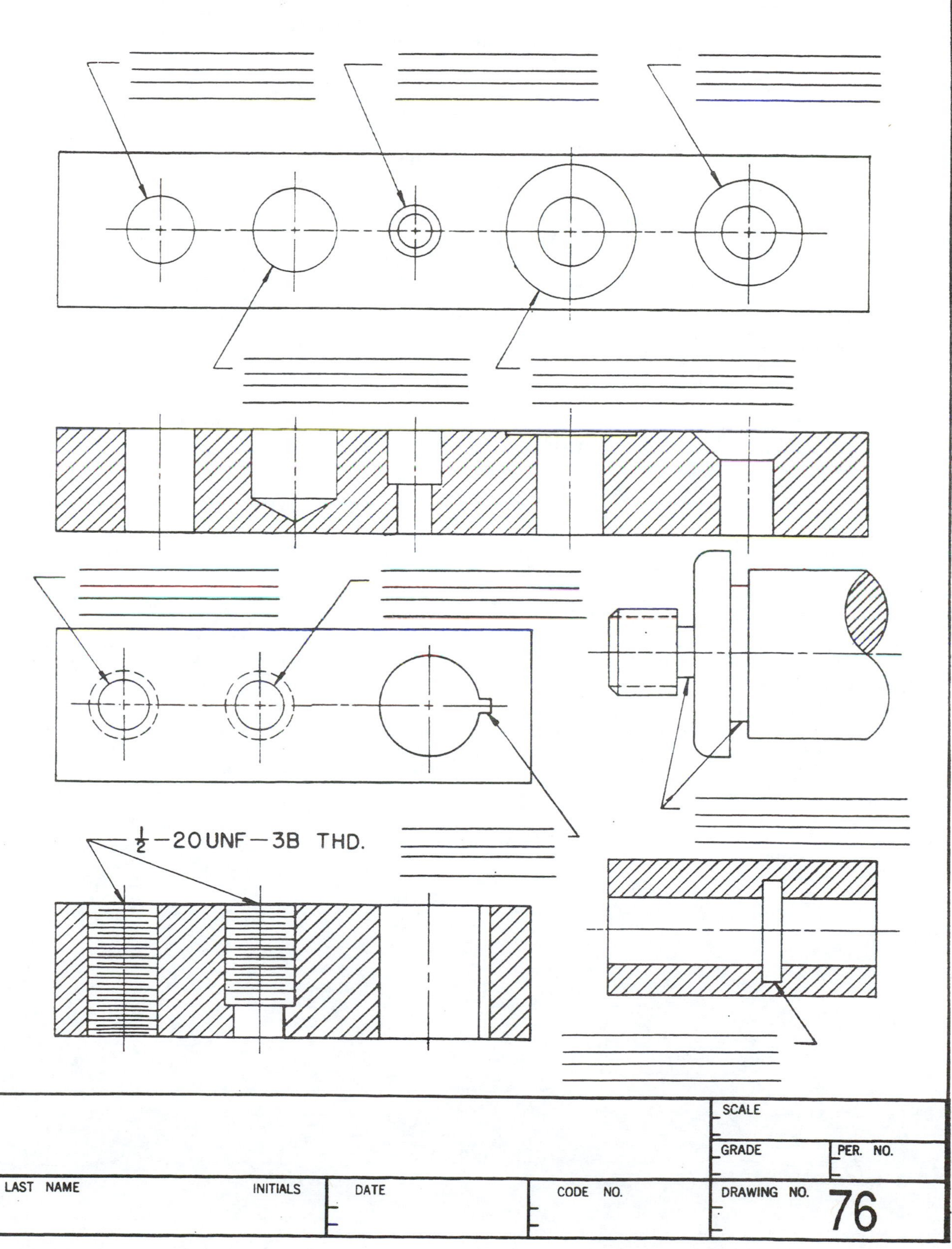

			SCALE	
			GRADE	PER. NO.
LAST NAME INITIALS	DATE	CODE NO.	DRAWING NO.	76

SPECIFICATION OF FASTENERS

Given below are several types of fasteners which must be properly specified in a bill of material or on order forms. Letter the correct (full not abridged) specification for each fastener using the guide lines provided. Needed dimensions may be obtained by scaling the drawings.

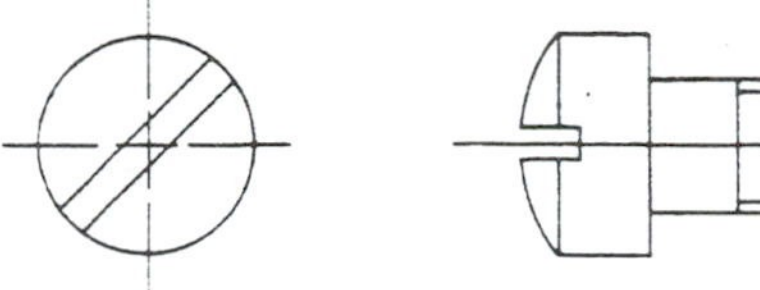

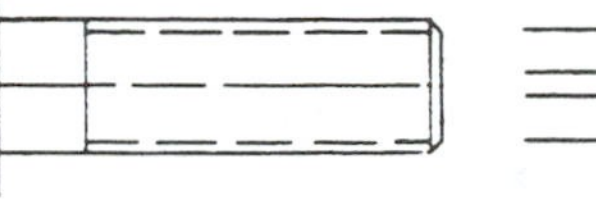

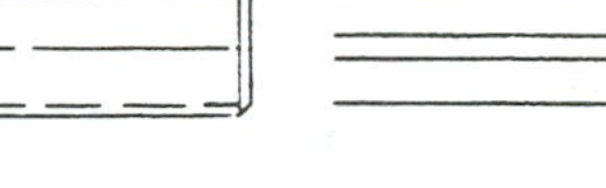

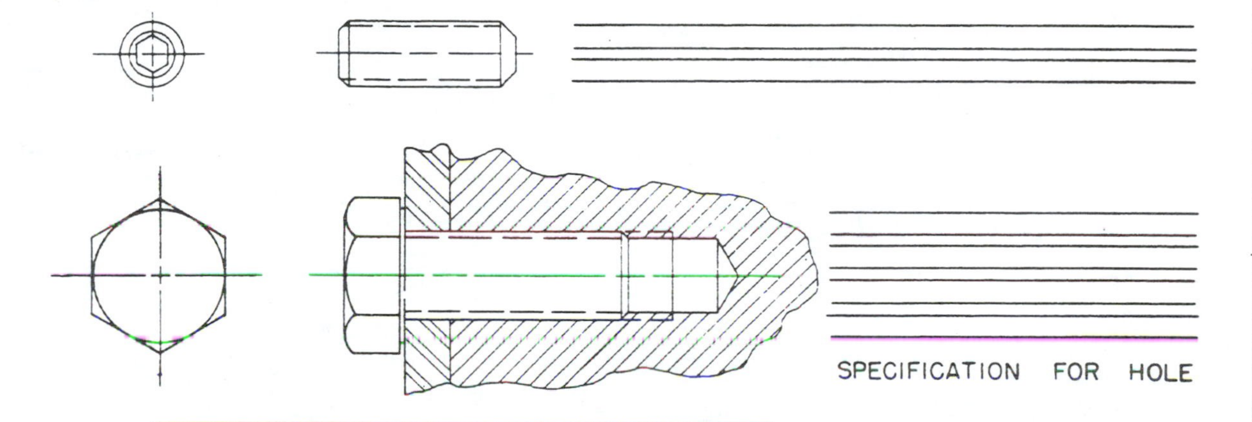

SPECIFICATION FOR HOLE

SPECIFICATION FOR FASTENER

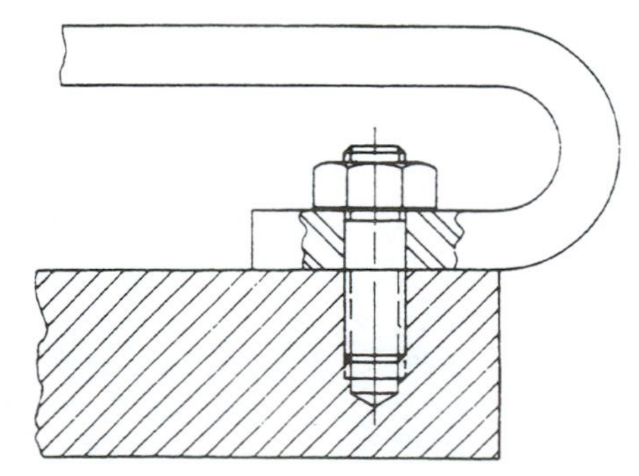

	SCALE	
	GRADE	PER. NO.
LAST NAME INITIALS	DATE	CODE NO.	DRAWING NO. 77

INTERSECTIONS

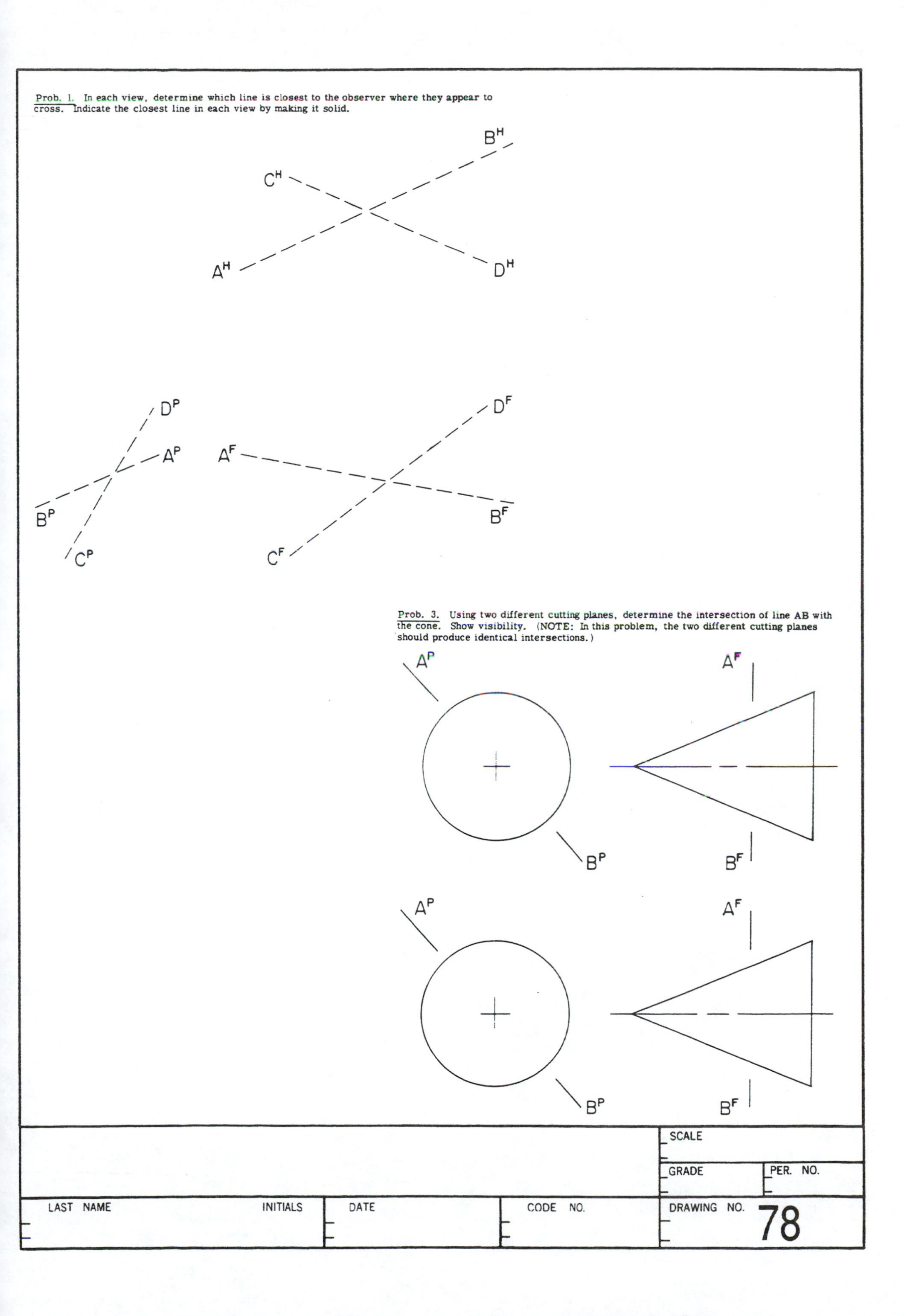
Prob. 1. In each view, determine which line is closest to the observer where they appear to cross. Indicate the closest line in each view by making it solid.
B^H
C^H
A^H
D^H
D^P
A^P
B^P
C^P
A^F
D^F
B^F
C^F
Prob. 3. Using two different cutting planes, determine the intersection of line AB with the cone. Show visibility. (NOTE: In this problem, the two different cutting planes should produce identical intersections.)
A^P
A^F
B^P
B^F
A^P
A^F
B^P
B^F
SCALE
GRADE
PER. NO.
LAST NAME
INITIALS
DATE
CODE NO.
DRAWING NO.
78

Prob. 5. Find the intersection of line XY with the sphere. Show visibility.

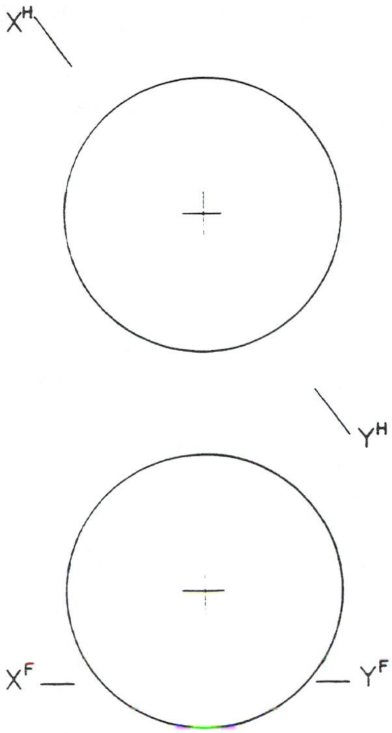

X^H

Y^H

X^F

Y^F

	SCALE	
	GRADE	PER. NO.

LAST NAME INITIALS	DATE	CODE NO.	DRAWING NO. 79

Prob. 7. Find the intersection of line PQ with the cylinder. (Note: The cylinder is of circular cross section.)

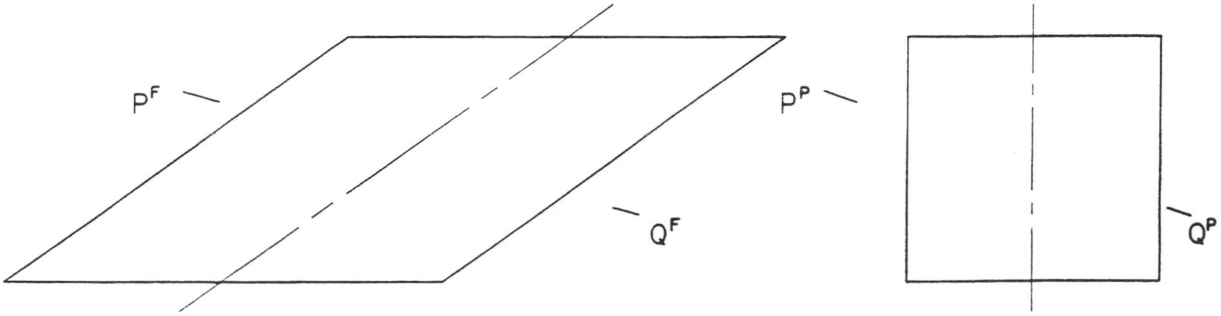

P^F

Q^F

P^P

Q^P

SCALE	
GRADE	PER. NO.

LAST NAME	INITIALS	DATE	CODE NO.	DRAWING NO. 80

Prob. 1. Find the intersection of the cone and the plane P_1 P_2 shown as an edge.

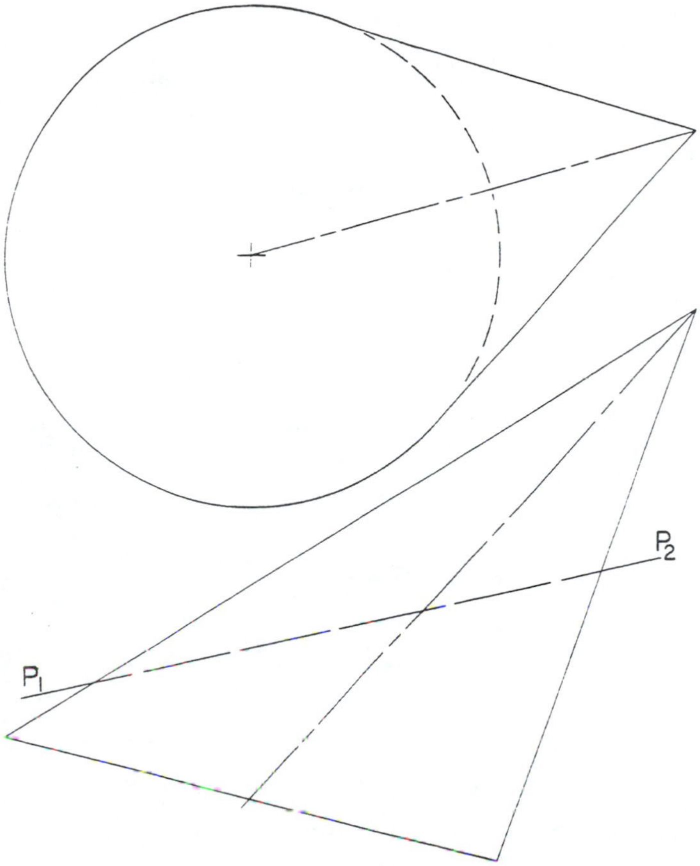

P_2

P_1

	SCALE	
	GRADE	PER. NO.

LAST NAME	INITIALS	DATE	CODE NO.	DRAWING NO. 82

PROBLEM SOLVING

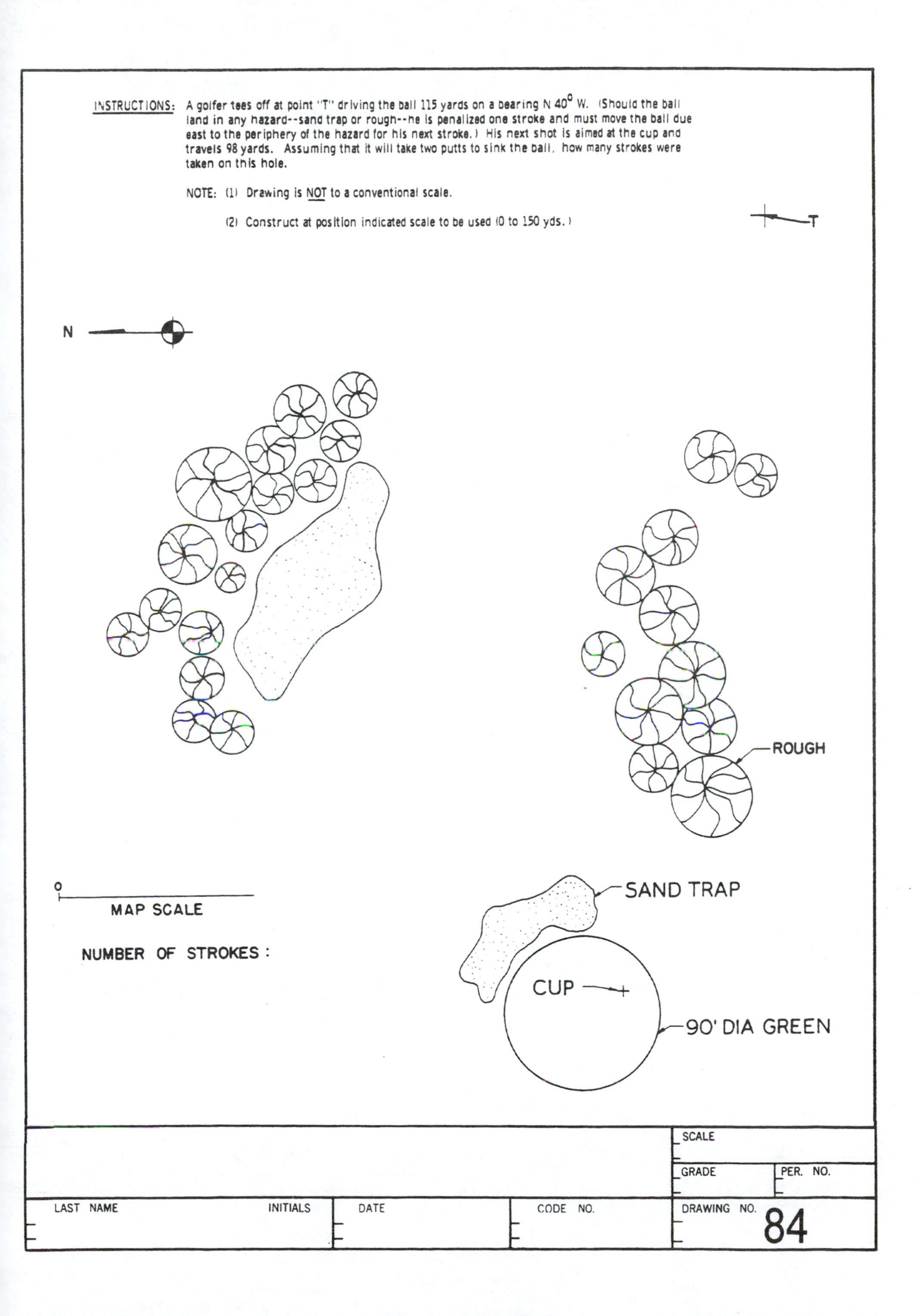
INSTRUCTIONS: A golfer tees off at point "T" driving the ball 115 yards on a bearing N 40° W. (Should the ball land in any hazard--sand trap or rough--he is penalized one stroke and must move the ball due east to the periphery of the hazard for his next stroke.) His next shot is aimed at the cup and travels 98 yards. Assuming that it will take two putts to sink the ball, how many strokes were taken on this hole.
NOTE: (1) Drawing is NOT to a conventional scale.
(2) Construct at position indicated scale to be used (0 to 150 yds.)
T
N
ROUGH
0
MAP SCALE
NUMBER OF STROKES :
SAND TRAP
CUP
90' DIA GREEN
SCALE
GRADE
PER. NO.
LAST NAME
INITIALS
DATE
CODE NO.
DRAWING NO.
84

SCALE: 1 in. = 60 ft.

INSTRUCTIONS:

a) Starting at point A, reconstruct the boundary lines of the plot of land owned by A. G. White Co. Point B is an existing corner stone.
b) Determine the bearing and length of the boundary line of the property which is parallel to the center line of the railroad right-of-way. Record the values along the boundary line.
c) Omit all lettering. Do not show angle values or boundary lengths except as called for in (b).
d) Show centerlines of right-of-ways.
e) Show all construction.

NOTE: A bearing angle of a property line in map survey is given from either North or South, the number of degrees East or West. S27°W is read: The line bears 27 degrees west of south. This is equivalent to N27°E which is read: The line bears 27 degrees east of north. The symbol, ℄, is read: Center line.

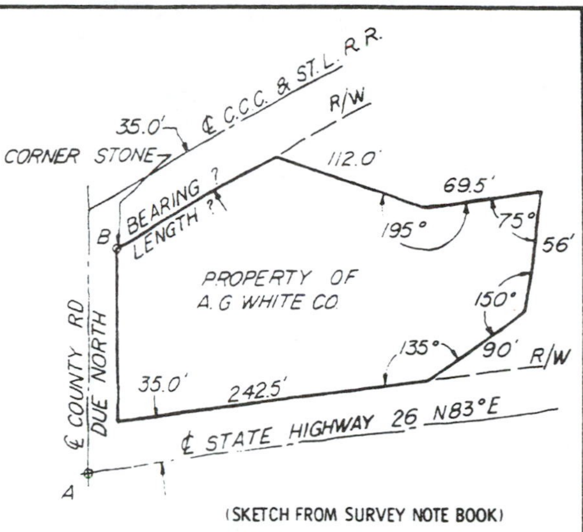

(SKETCH FROM SURVEY NOTE BOOK)

NOTE: Sketches are not drawn to scale. The width of a roadway is measured perpendicular to the center line.

B

N

A

	SCALE	
	GRADE	PER. NO.
LAST NAME / INITIALS / DATE / CODE NO.	DRAWING NO. 85	

SCALE: 1 in. = 2 ft.

INSTRUCTIONS: A chemical solution tank is described as follows: The base is 3.00' x 4.05', the tank is 7.0' high, the trapezoidal cross section taken at a right angle to the 4.05' dimension shows the right and left side walls inclined upward and outward at angles of 60° and 75° to the base respectively. When the tank is filled with 1000 gallons of chemical solution, the surface of the fluid is 6.0' above the bottom of the tank.

1) Complete the cross section ABCD (shown shaded on pictorial) of the tank. Show all constructions.

2) Construct a volume scale on the cross section in such a way that every 200 gallons of solution can be read directly. Label the graduations within the boundaries of the tank. Do not dimension.

3) Measure and record the distance along the sides of the tank from the bottom to the 1000 gallon mark.

60° SIDE =

75° SIDE =

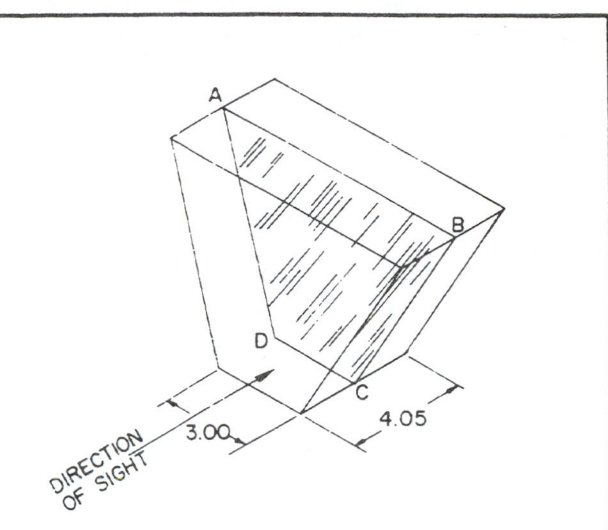

STARTING LINE

LAST NAME	INITIALS	DATE	CODE NO.	SCALE
				GRADE / PER. NO.
				DRAWING NO. 86

Prob. 1. A contractor who builds outdoor swimming pools wants to deviate from a conventional rectangular form. The rectangular shape, as shown, is retained in the new design as one-half of the pool. The other half of the new design is to be elliptical. The diving board is not on the center of the pool. Construct the elliptical portion of the new design using the concentric circle method. Show all construction.

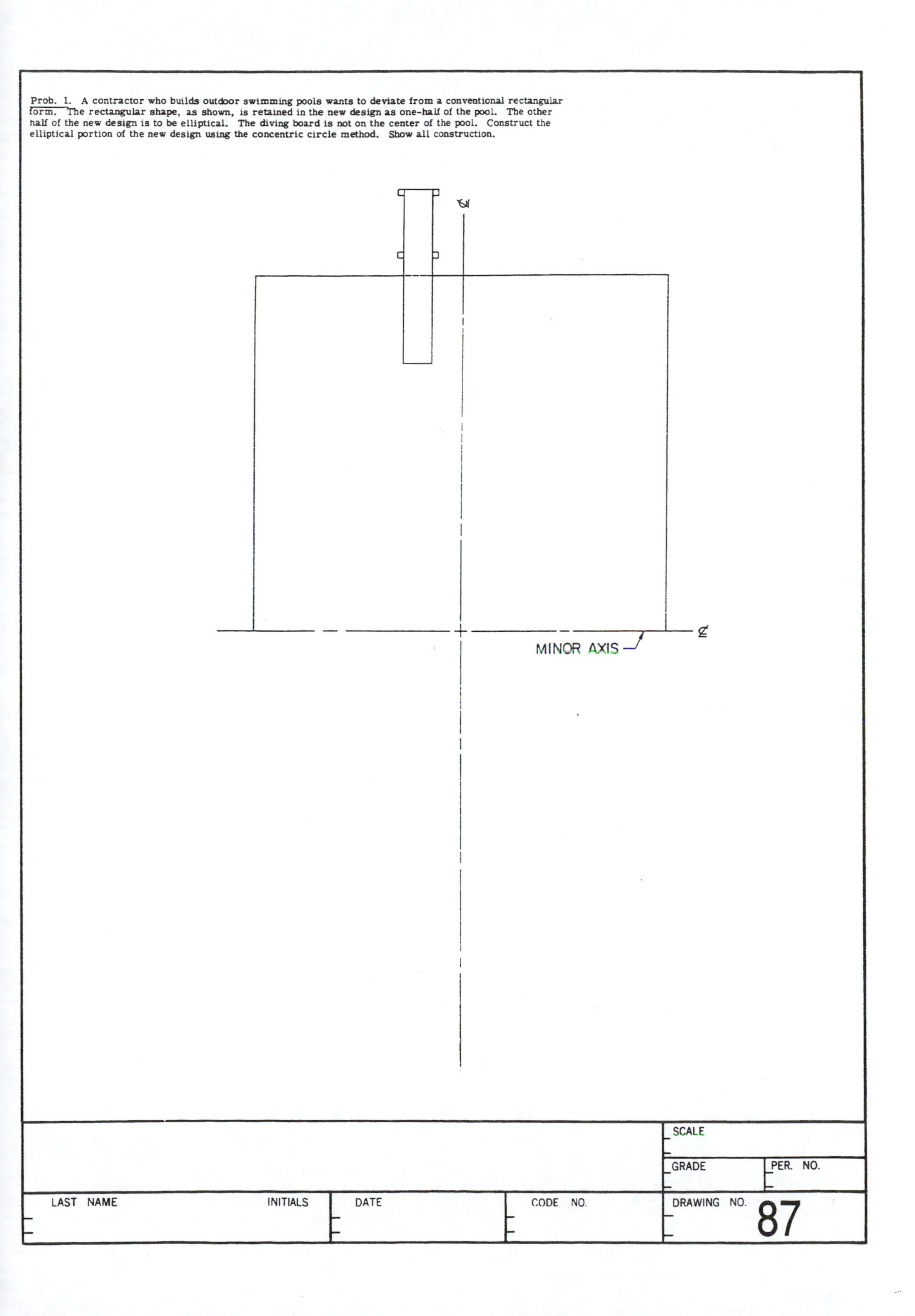

Prob. 2. It is desired to construct the elliptical lunar trajectories of an unmanned lunar spacecraft after it has been hurled into space along a path that is tangent to the given line at point P by an Atlas-Agena D booster. When it reaches point P its control rockets are fired in such a manner as to place the spacecraft in an elliptical orbit which has the center of the moon as one focus and the major axis in the direction as shown by the centerline. After exactly one and one-half orbits the control rockets are fired again so as to place the spacecraft in a smaller elliptical orbit which also has the center of the moon as one focus. The major axis of the smaller orbit makes an angle of 30° counter-clockwise with the major diameter of the first orbit. Plot the first orbit using the foci-to-foci method and plot the smaller second orbit by the trammel method. Determine the location of the second firing.

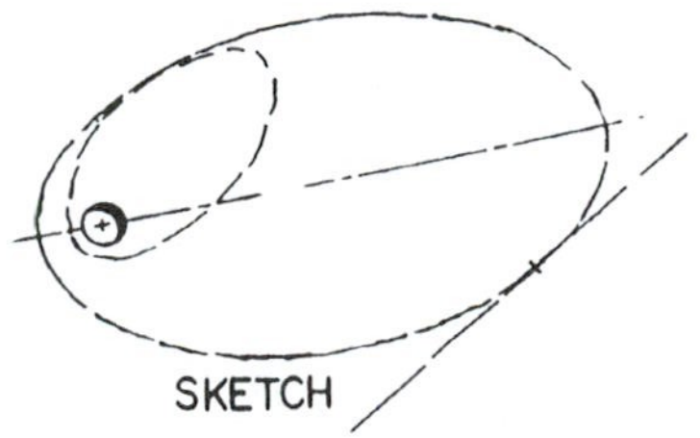

SKETCH

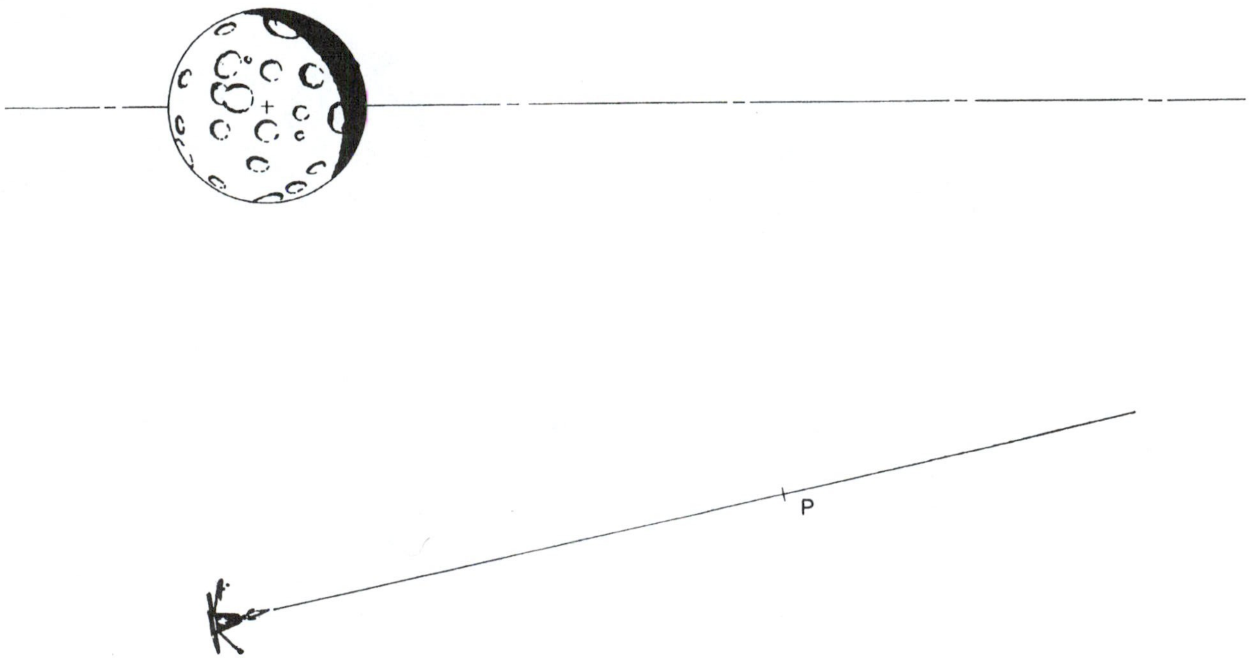

P

LAST NAME	INITIALS	DATE	CODE NO.	SCALE	
				GRADE	PER. NO.
				DRAWING NO.	88

3-D MODELING

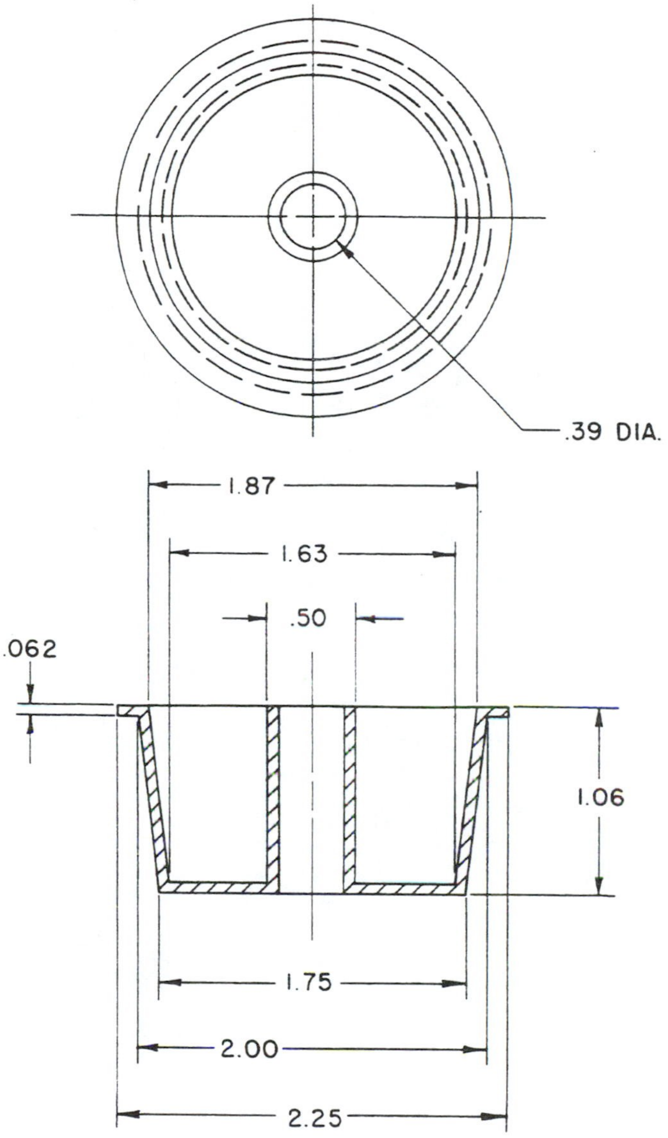

.39 DIA.

1.87

1.63

.50

.062

1.06

1.75

2.00

2.25

NUMBER REQ: 1	MATERIAL: POLYSTYRENE	FINISH:		
UNLESS OTHERWISE SPECIFIED DIMENSIONS ARE IN INCHES. TOLERANCES ARE:	WEIGHT: .01 LB	NAME	DATE	TITLE CONE- LEAD POINTER
FRACTIONS = ±				
.xx DECIMALS = ± .01	DRAWN BY:			
.xxx DECIMALS = ± .002	CHECKED BY:			
ANGLES: ± .5°	APPROVED BY:			REFERENCE DRAWING:
x.x METRIC = ±				
x.xx METRIC = ±	SCALE: 1 = 1		A	DRAWING NUMBER 89

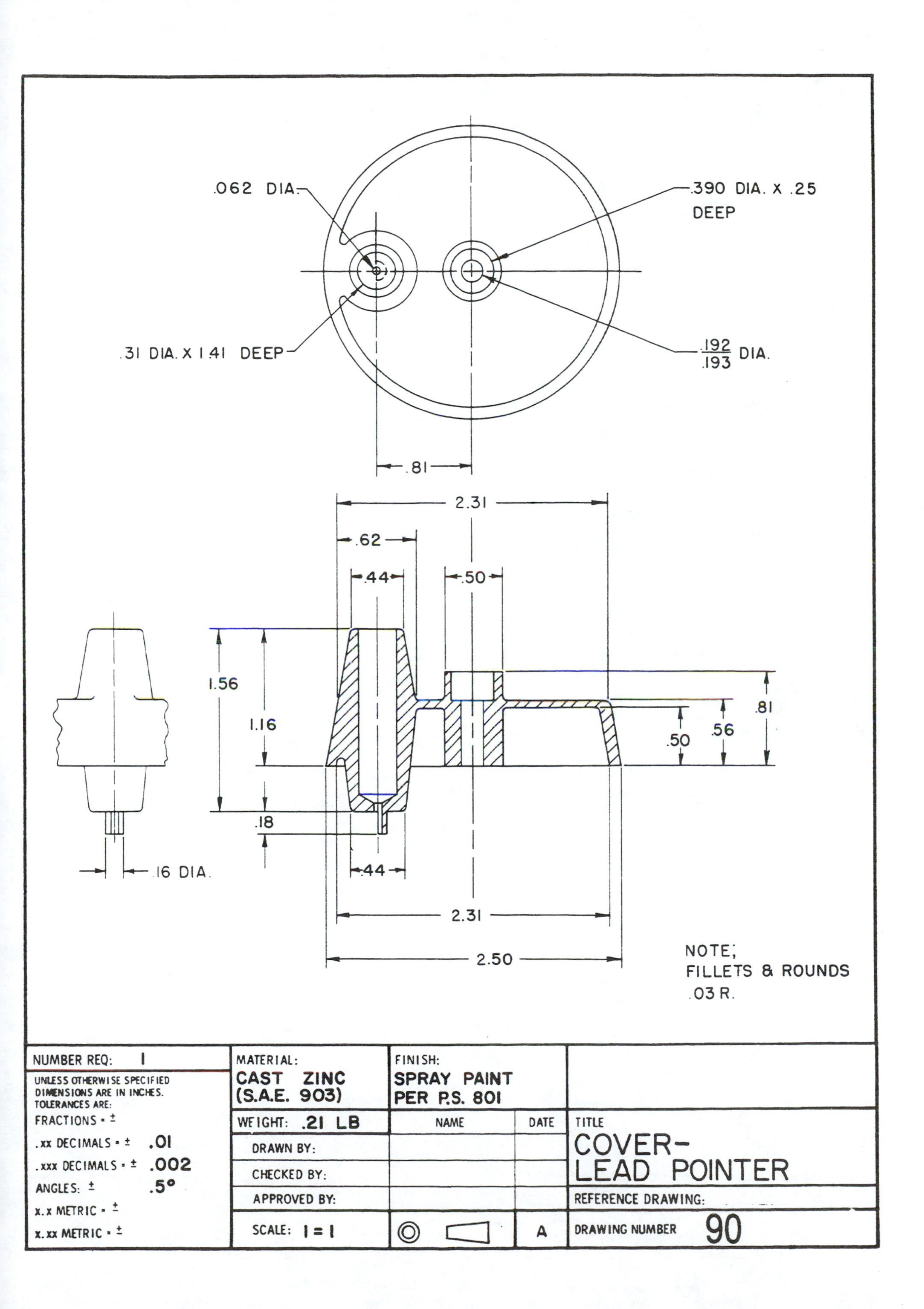
.062 DIA.
.390 DIA. X .25 DEEP
.31 DIA. X 1.41 DEEP
.192/.193 DIA.
.81
2.31
.62
.44
.50
1.56
1.16
.18
.44
.16 DIA.
.81
.50
.56
2.31
2.50
NOTE;
FILLETS & ROUNDS .03 R.
NUMBER REQ: 1
UNLESS OTHERWISE SPECIFIED DIMENSIONS ARE IN INCHES. TOLERANCES ARE:
FRACTIONS = ±
.xx DECIMALS = ± .01
.xxx DECIMALS = ± .002
ANGLES: ± .5°
x.x METRIC = ±
x.xx METRIC = ±
MATERIAL:
CAST ZINC (S.A.E. 903)
FINISH:
SPRAY PAINT PER P.S. 801
WEIGHT: .21 LB
NAME
DATE
DRAWN BY:
CHECKED BY:
APPROVED BY:
SCALE: 1 = 1
A
TITLE
COVER-
LEAD POINTER
REFERENCE DRAWING:
DRAWING NUMBER 90

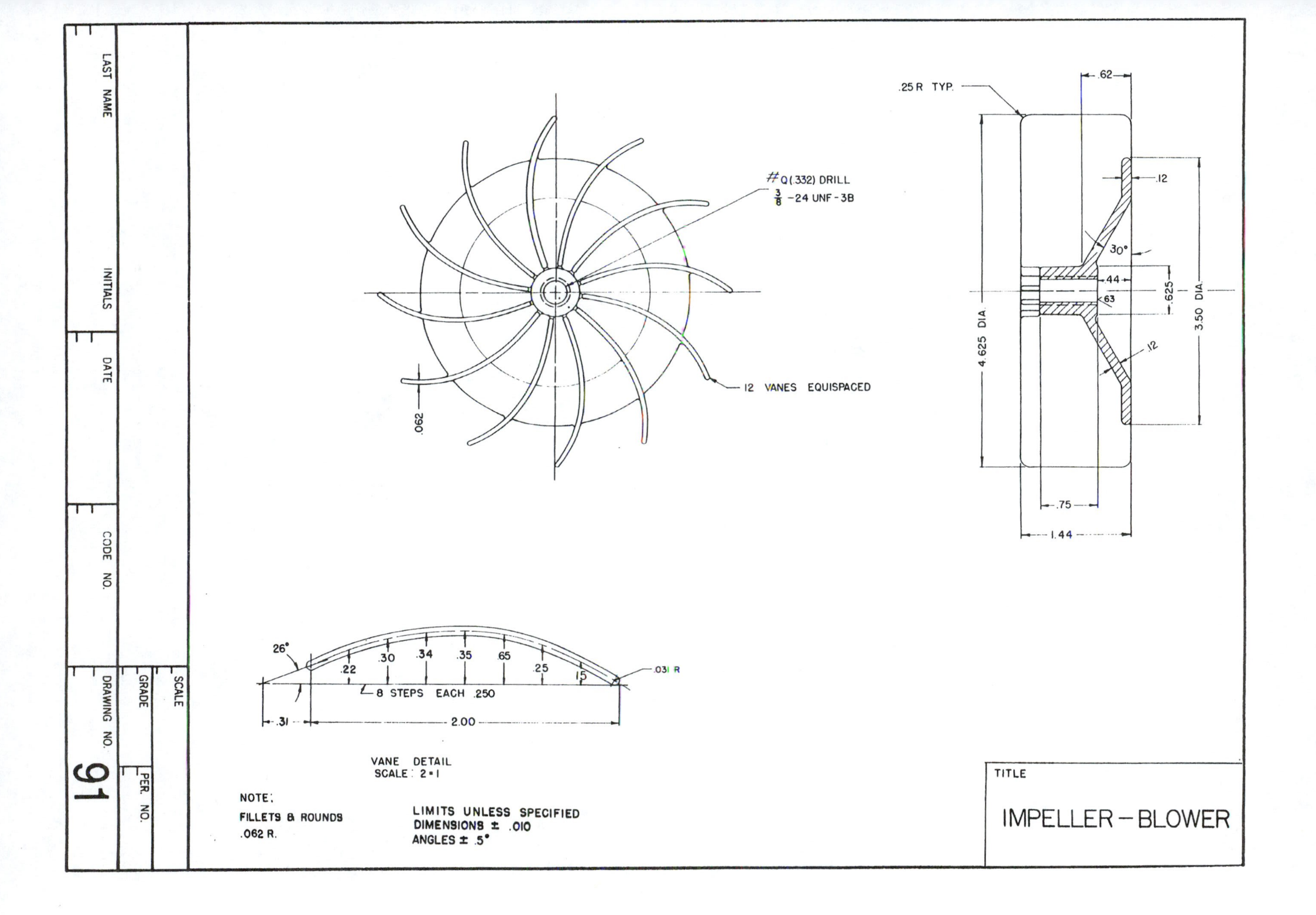

Q (.332) DRILL
3/8 –24 UNF – 3B
12 VANES EQUISPACED
.062
.25 R TYP.
.62
.12
30°
.44
.63
.625
3.50 DIA.
4.625 DIA.
.12
.75
1.44
26°
.22
.30
.34
.35
.65
.25
.15
.031 R
8 STEPS EACH .250
.31
2.00
VANE DETAIL
SCALE : 2 = 1
NOTE:
FILLETS & ROUNDS
.062 R.
LIMITS UNLESS SPECIFIED
DIMENSIONS ± .010
ANGLES ± .5°
TITLE
IMPELLER – BLOWER
LAST NAME
INITIALS
DATE
CODE NO.
DRAWING NO.
91
GRADE
PER. NO.
SCALE

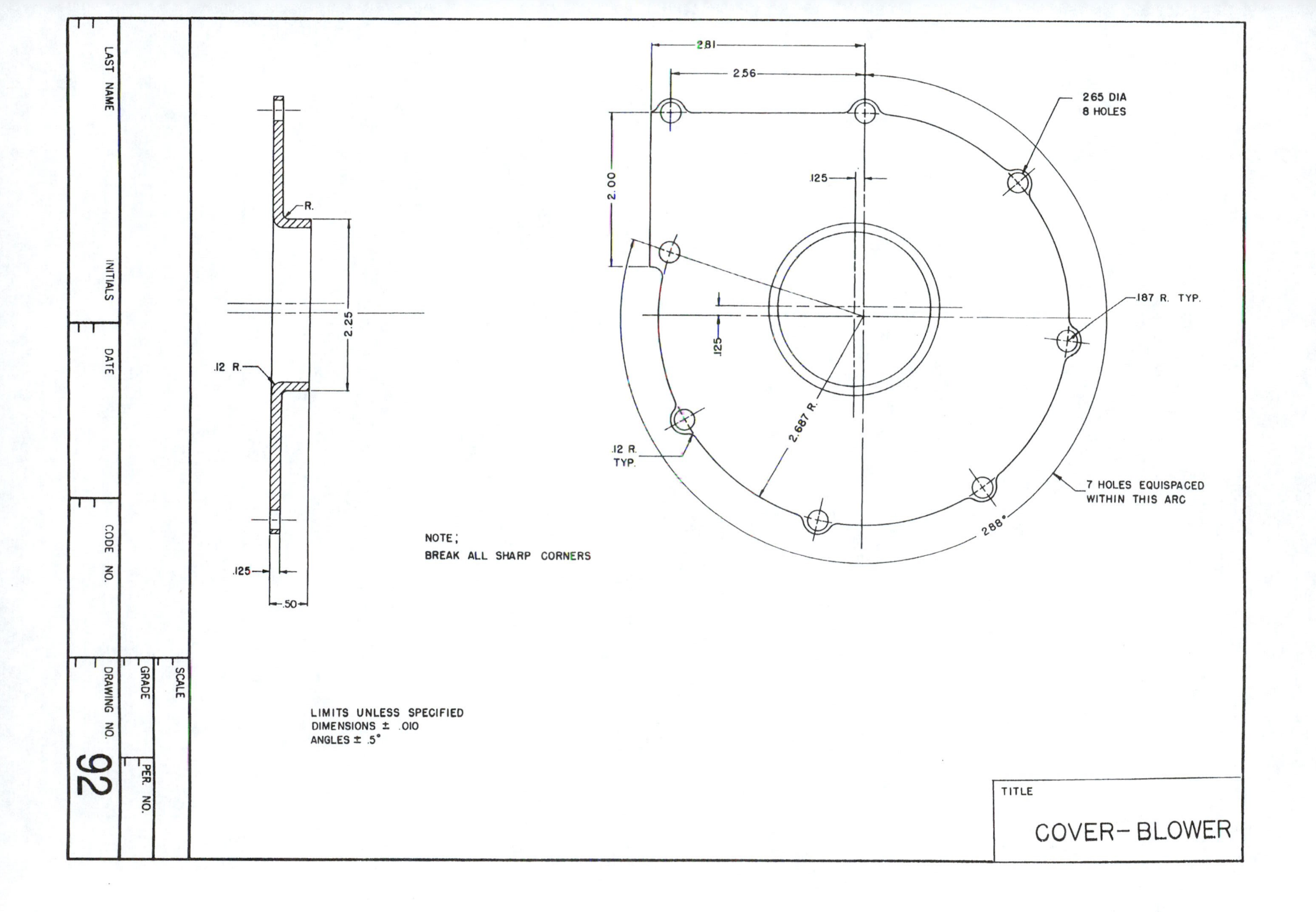

2.81
2.56
.265 DIA
8 HOLES
.125
2.00
R.
2.25
.187 R. TYP.
.125
.12 R.
2.687 R.
.12 R.
TYP.
7 HOLES EQUISPACED
WITHIN THIS ARC
288°
NOTE;
BREAK ALL SHARP CORNERS
.125
.50
LIMITS UNLESS SPECIFIED
DIMENSIONS ± .010
ANGLES ± .5°
TITLE
COVER- BLOWER
LAST NAME
INITIALS
DATE
CODE NO.
DRAWING NO.
92
GRADE
PER. NO.
SCALE

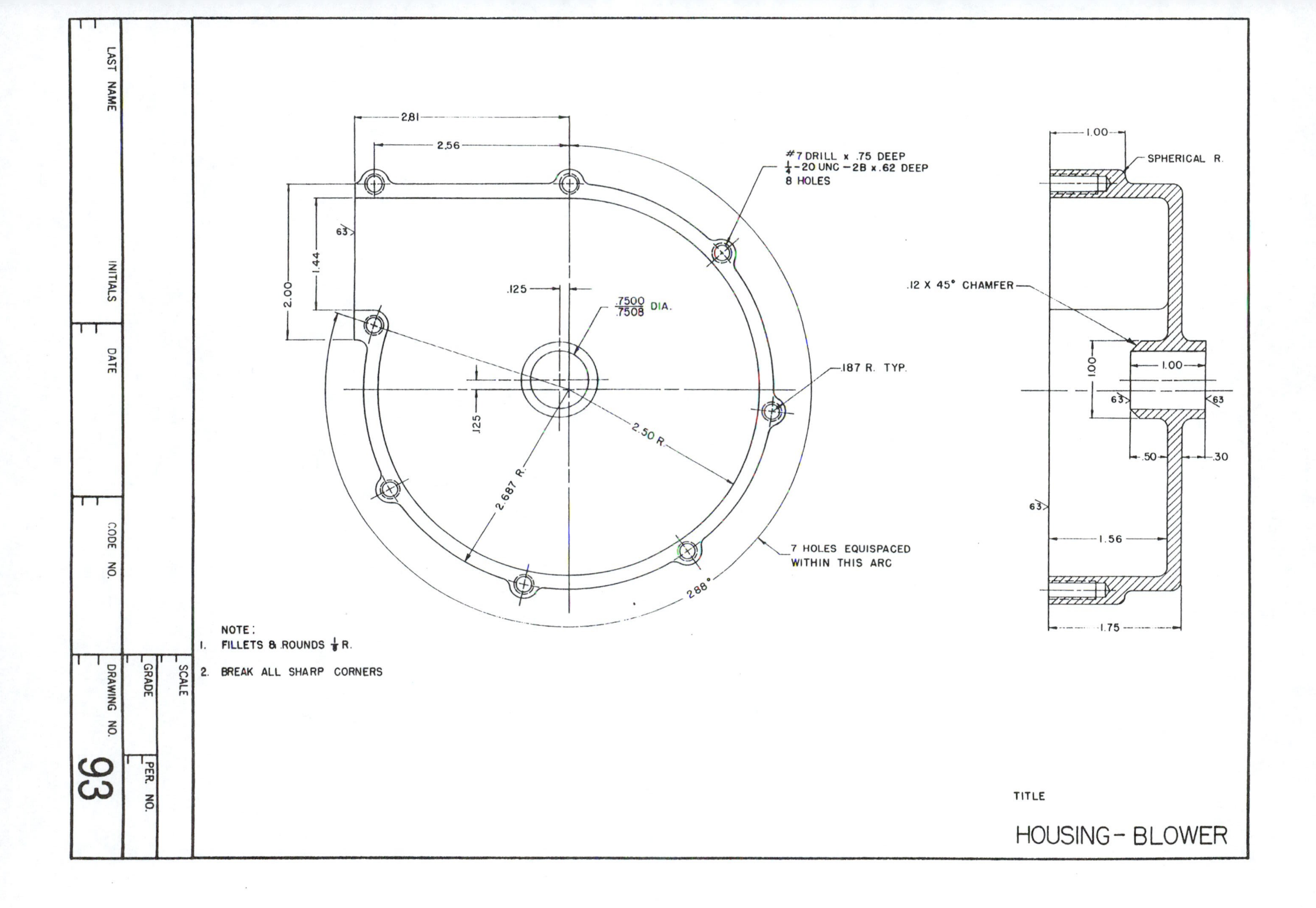
2.81
2.56
2.00
1.44
.125
.125
.7500/.7508 DIA.
2.50 R.
2.687 R.
63
#7 DRILL x .75 DEEP
1/4-20 UNC-2B x .62 DEEP
8 HOLES
.187 R. TYP.
7 HOLES EQUISPACED
WITHIN THIS ARC
288°
1.00
SPHERICAL R.
.12 X 45° CHAMFER
1.00
1.00
.50
.30
1.56
1.75
NOTE:
1. FILLETS & ROUNDS 1/8 R.
2. BREAK ALL SHARP CORNERS
TITLE
HOUSING - BLOWER
LAST NAME
INITIALS
DATE
CODE NO.
SCALE
GRADE
PER. NO.
DRAWING NO.
93

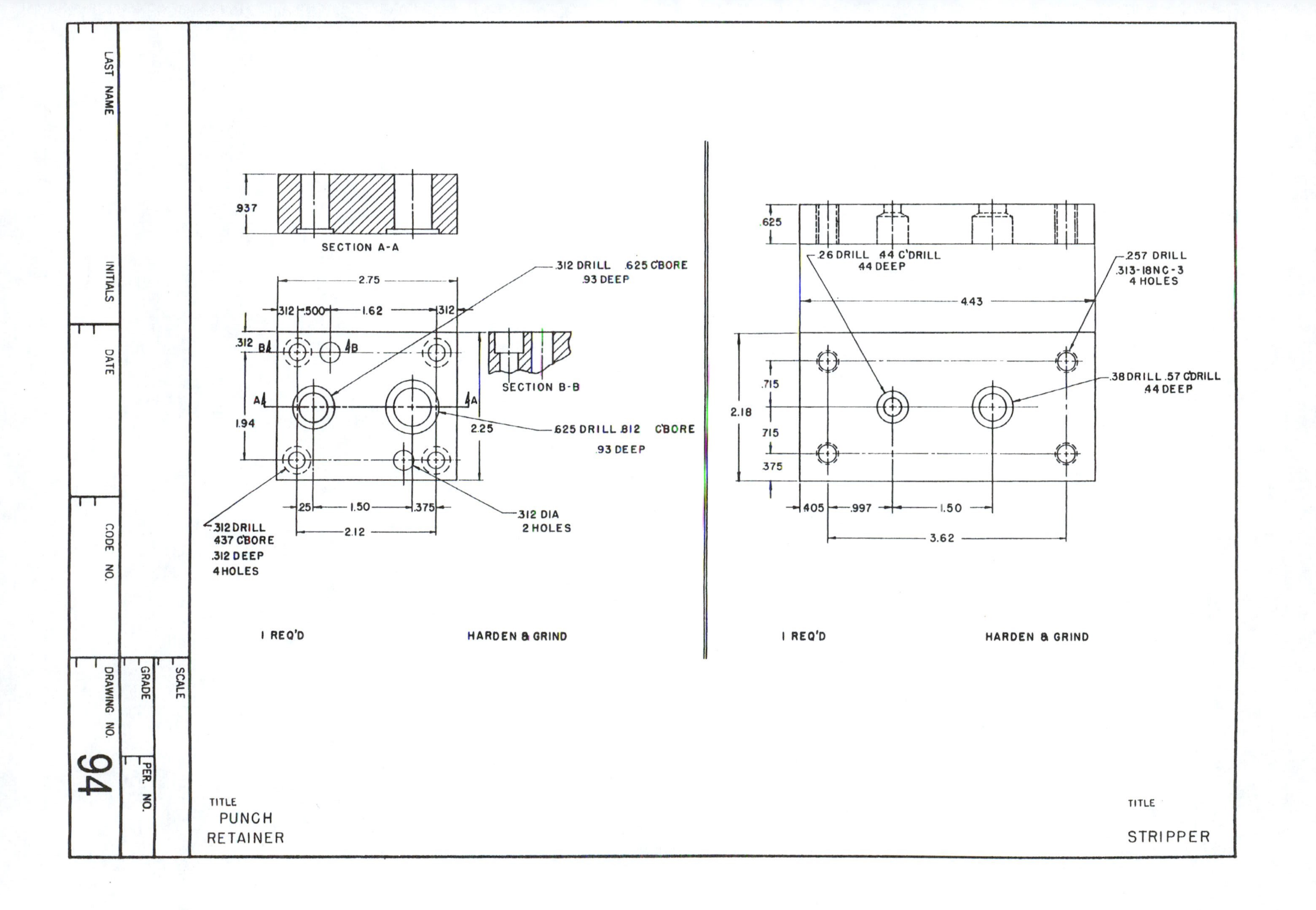
.937
SECTION A-A
.312 DRILL .625 C'BORE
.93 DEEP
2.75
.312
.500
1.62
.312
.312
SECTION B-B
1.94
2.25
.625 DRILL .812 C'BORE
.93 DEEP
.25
1.50
.375
2.12
.312 DIA
2 HOLES
.312 DRILL
.437 C'BORE
.312 DEEP
4 HOLES
1 REQ'D
HARDEN & GRIND
TITLE
PUNCH
RETAINER
.625
.26 DRILL .44 C'DRILL
.44 DEEP
.257 DRILL
.313-18NC-3
4 HOLES
4.43
.38 DRILL .57 C'DRILL
.44 DEEP
.715
2.18
.715
.375
.405
.997
1.50
3.62
1 REQ'D
HARDEN & GRIND
TITLE
STRIPPER
LAST NAME
INITIALS
DATE
CODE NO.
DRAWING NO.
94
GRADE
PER. NO.
SCALE

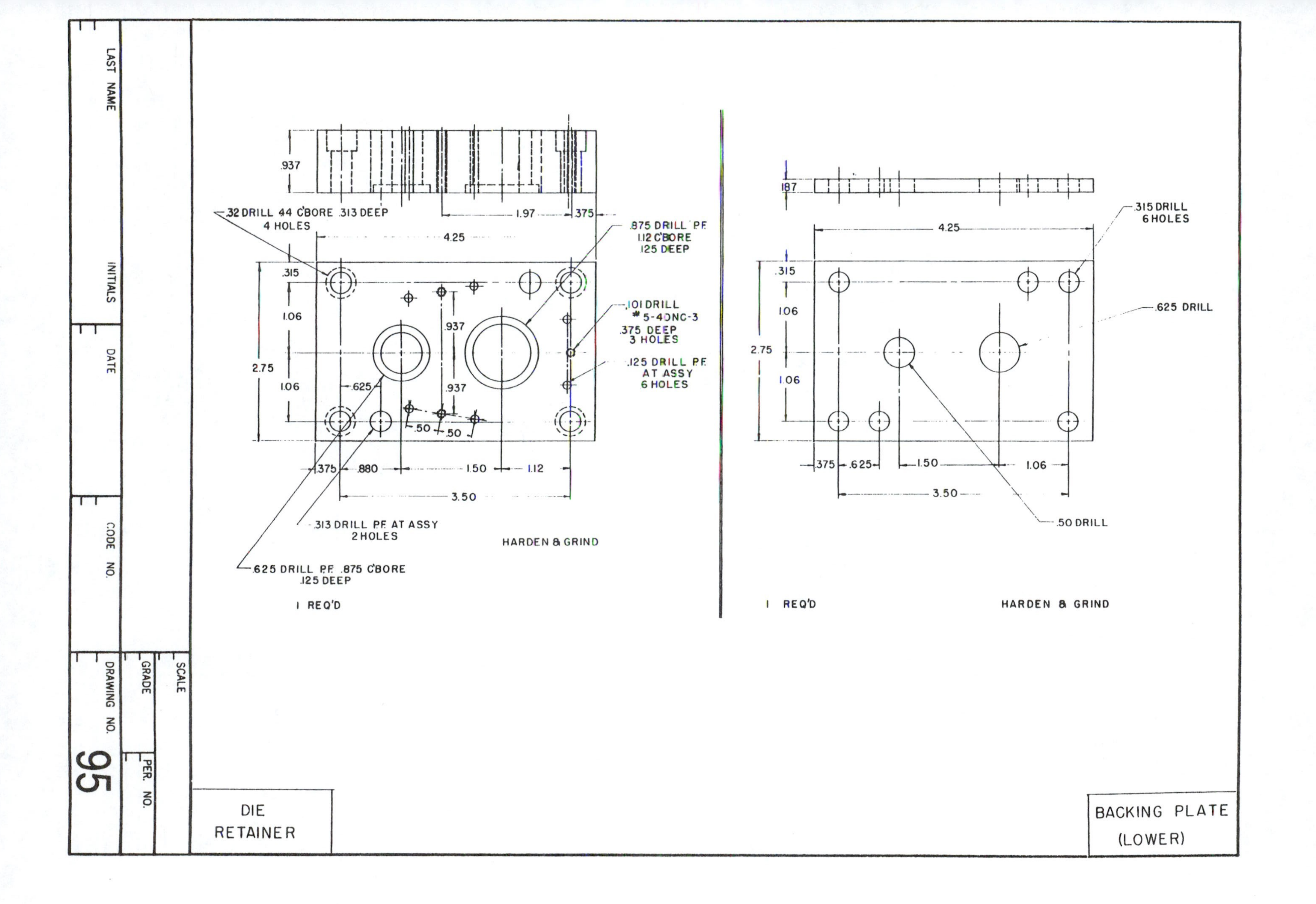
.32 DRILL .44 C'BORE .313 DEEP
4 HOLES
.875 DRILL P.F.
1.12 C'BORE
.125 DEEP
.101 DRILL
#5-40NC-3
.375 DEEP
3 HOLES
.125 DRILL P.F.
AT ASSY
6 HOLES
.313 DRILL P.F. AT ASSY
2 HOLES
HARDEN & GRIND
.625 DRILL P.F. .875 C'BORE
.125 DEEP
1 REQ'D
.315 DRILL
6 HOLES
.625 DRILL
.50 DRILL
1 REQ'D
HARDEN & GRIND
DIE
RETAINER
BACKING PLATE
(LOWER)
LAST NAME
INITIALS
DATE
CODE NO.
DRAWING NO.
95
GRADE
PER. NO.
SCALE

FORMS

FORM SA

	SCALE	
	GRADE	PER. NO.

LAST NAME INITIALS	DATE	CODE NO.	DRAWING NO. 96

FORM SA

	SCALE	
	GRADE	PER. NO.

LAST NAME INITIALS	DATE	CODE NO.	DRAWING NO. 97

FORM SA

	SCALE	
	GRADE	PER. NO.

LAST NAME INITIALS	DATE	CODE NO.	DRAWING NO. 98

FORM SA

	SCALE	
	GRADE	PER. NO.

LAST NAME INITIALS	DATE	CODE NO.	DRAWING NO. 99

FORM SA

	SCALE	
	GRADE	PER. NO.

LAST NAME INITIALS	DATE	CODE NO.	DRAWING NO. 100

FORM SA

	SCALE	
	GRADE	PER. NO.
LAST NAME INITIALS	DATE	CODE NO.	DRAWING NO. 101

NUMBER REQ:	MATERIAL:	FINISH:		
UNLESS OTHERWISE SPECIFIED DIMENSIONS ARE IN INCHES. TOLERANCES ARE:				
FRACTIONS = ±	WEIGHT:	NAME	DATE	TITLE
.xx DECIMALS = ±	DRAWN BY:			
.xxx DECIMALS = ±	CHECKED BY:			
ANGLES: ±	APPROVED BY:			REFERENCE DRAWING:
x.x METRIC = ±	SCALE:	◎	A	DRAWING NUMBER 102
x.xx METRIC = ±				

NUMBER REQ:	MATERIAL:	FINISH:		
UNLESS OTHERWISE SPECIFIED DIMENSIONS ARE IN INCHES. TOLERANCES ARE:				
FRACTIONS = ±	WEIGHT:	NAME	DATE	TITLE
.xx DECIMALS = ±	DRAWN BY:			
.xxx DECIMALS = ±	CHECKED BY:			
ANGLES: ±	APPROVED BY:			REFERENCE DRAWING:
x.x METRIC = ±	SCALE:	◎	A	DRAWING NUMBER 103
x.xx METRIC = ±				

		SCALE	
		GRADE	PER. NO.
LAST NAME INITIALS	DATE	CODE NO.	DRAWING NO. 104

	SCALE	
	GRADE	PER. NO.

LAST NAME	INITIALS	DATE	CODE NO.	DRAWING NO. 105

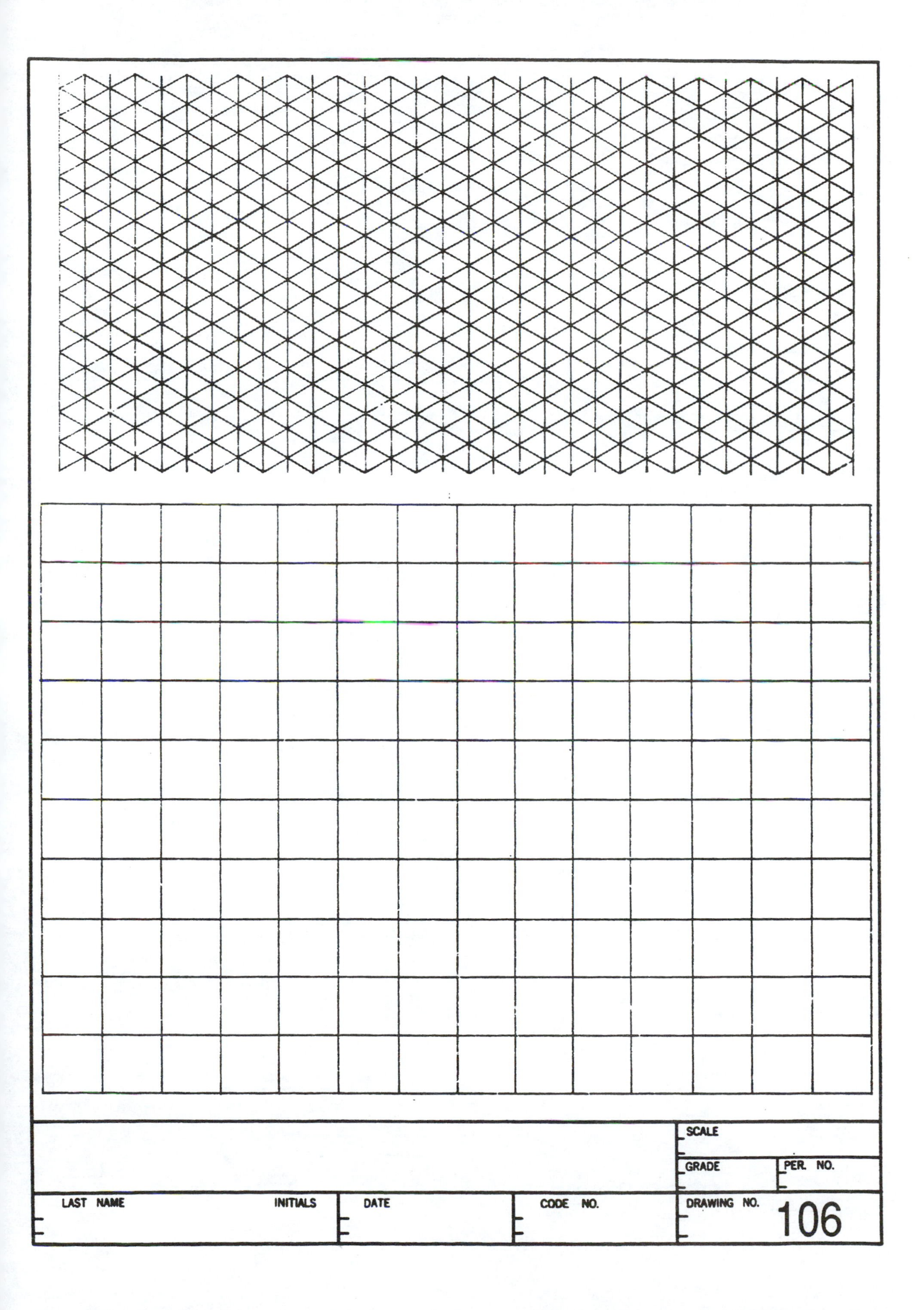
SCALE
GRADE
PER. NO.
LAST NAME
INITIALS
DATE
CODE NO.
DRAWING NO.
106

		SCALE	
		GRADE	PER. NO.
LAST NAME INITIALS	DATE	CODE NO.	DRAWING NO. 107

	SCALE	
	GRADE	PER. NO.

LAST NAME INITIALS	DATE	CODE NO.	DRAWING NO. 108

SCALE	
GRADE	PER. NO.

LAST NAME INITIALS	DATE	CODE NO.	DRAWING NO. 109

	SCALE	
	GRADE	PER. NO.
LAST NAME INITIALS	DATE	CODE NO.	DRAWING NO. 110

	SCALE	
	GRADE	PER. NO.

LAST NAME INITIALS	DATE	CODE NO.	DRAWING NO. 111

SCALE	
GRADE	PER. NO.

LAST NAME	INITIALS	DATE	CODE NO.	DRAWING NO. 112